Set Theory–
An Operational Approach

Set Theory–
An Operational Approach

Luis E. Sanchis

Professor Emeritus
Syracuse University
Syracuse, New York

CRC Press
Taylor & Francis Group
Boca Raton London New York

CRC Press is an imprint of the
Taylor & Francis Group, an **informa** business

First published 1996 by Gordon and Breach Science Publishers

Published 2021 by CRC Press
Taylor & Francis Group
6000 Broken Sound Parkway NW, Suite 300
Boca Raton, FL 33487-2742

ISBN 13: 978-90-5699-507-2 (hbk)

DOI: 10.1201/9780203739860

British Library Cataloguing in Publication Data

Sanchis, Luis E.
Set theory—an operational approach
1.Set theory
I.Title
511.3'22

To Gabriela and Laura
(in alphabetical order)

TOINETTE: Je suis médecin passager, qui vais de ville en ville, de province en province, de royaume en royaume, pour chercher d'illustres matières à ma capacité, pour trouver des malades dignes de m'occuper, capables d'exercer les grands et beaux secrets que j'ai trouvés dans la mèdicine. Je dédaigne de m'amuser à ce menu fatras de maladies ordinaires, à ces bagatelles de rhumatismes et de fluxions, à ces fiévrotes, à ces vapeurs, et à ces migraines. Je veux des maladies d'importances, de bonnes fièvres continues, avec des transports au cerveau, de bonnes fièvres pourprées, de bonnes pestes, de bonnes hydropisies formées, de bonnes pleurésies, avec des inflammations de poitrine, c'est là que je me plais, c'est là que je triomphe; et je voudrais, monsieur, que vous eussiez toutes les maladies que je viens de dire, que vous fussiez abandonné de tous les médecins, désespéré, à l'agonie, pour vous montrer l'excellence de mes remèdes, et l'envie que j'aurais de vous rendre service.

—Molière
Le Malade Imaginaire

Contents

Preface

For most of this century, set theory has been identified with the study of special systems formalized in the framework of first-order logic (and in some cases a weak second-order logic). The prevalent system has been the well-known Zermelo–Fraenkel theory, often including the axiom of choice. These systems are relatively simple, being partially recursively decidable, and can be shown to contain a great deal of current mathematical practice. They are assumed to be consistent.

The formalization of set theory is a somewhat late occurrence, for the initial attempts by Cantor and Dedekind were organized around a general conception where sets are quasi-real mathematical entities and the universe of sets a kind of absolute model. Under such an approach the object of set theory was to prove true statements about the sets in the transcendental model. This approach collapsed with the advent of the paradoxes, which eventually led to the familiar formalization.

The introduction of formalization changed to a great extent the nature of the theory. In particular, the need to invoke explicitly the absolute model disappeared, for a formal system is a self-contained structure that can be studied independently of any particular interpretation. In many cases, the axiomatization of set theory was considered to be just an example of a general tendency in mathematics. In others, it was understood as a method to make notions that were intrinsically vague and ambiguous more precise.

In fact, the formalization of set theory allows mathematicians to operate in a systematic ambiguity, a situation that occurs frequently in the history of science. Questions about the validity of different hypotheses (in particular, the famous continuum hypothesis) are very often considered in terms of their possible truth or falsity, and this implicitly invokes a quasi-real universe of sets. I can go further and say that the general assumption that Zermelo–Fraenkel axioms are consistent originates in the implicit assumption that such axioms are true in some transcendental interpretation. On the other hand, when objections are raised about such an assumption, logicians retreat immediately to the classical formalistic position where axioms are simply formal hypotheses to be verified in particular interpretations. Since interpretations are always specified in terms of sets, this brings us back to the original situation. Still, practitioners will proudly proclaim that they simply derive conclusions from given axioms,which is the usual practice in mathematics. And they are certainly right.

It appears to me that, notwithstanding formal proclamations and the commitment to work in the frame of a fixed axiomatic system, it is the realistic conception that determines the actual meaning of set theory. It is sufficient to compare the particular manner in which assertions about sets are formulated in most presentations to realize that these assertions are intended to describe some special type of reality. On the other hand, standard axiomatic systems in mathematics are intended to describe a plurality of possible interpretations with the implicit understanding that such interpretations will be provided by set theory.

The conflict between formalism and realism pervades the actual practice of set theory, and this ambiguity is a clear indication that set theory cannot be reduced to an axiomatic system similar to those that occur in ordinary mathematics. Set theory is a very special type of mathematical theory and must be approached with a proper consideration of its peculiarities.

The point I try to make in this work is that set theory is concerned with some type of reality, only it is an ambiguous sort of reality. In particular, the universe of sets cannot be constructed as a complete collection where global quantification is legitimate. More precisely, I think that the role of set theory is to make explicit different possible constructions where sets are derived from given sets. At the same time we must recognize the fact that there is no actual construction for the universe of sets,and this situation must be taken into account in the formulation of the theory.

I express this distinction by means of a classical terminology in the foundations of mathematics. Constructions that involve the universe I call *impredicative*. As a consequence, in this work I am committed to constructions that are *predicative* (i.e., non-impredicative). So I take a large view of what is predicative, that goes beyond the usual practice. In particular, I am prepared to accept constructions where a set is introduced by a reference to a totality that contains the set, provided such a totality has been introduced by a legitimate (= predicative) construction. References to the universe are not predicative, because there is no predicative construction of the universe.

The crucial example where a non-trivial set is introduced by a predicative construction is the set ω of natural numbers, and I discuss in chapter 7 the details of the construction. There is no surprise here, and the reader will see that I rely on the obvious fundamental process that generates the natural numbers.

On the other hand, I object to the standard "definition" of the power-set of a set X, as the collection of all subsets of X. The reason is that this is an impredicative construction, since the potential sets to be selected by the property of being a subset of X are taken from the universe. Actually, the definition says that the elements of the power-set of X are those sets *in the universe* that are subsets of X. In chapter 8 I give a predicative construction of the power-set that involves the axiom of choice.

In this volume references to a complete universe are eliminated via a formal system consisting of set operation rules and set predicate rules, where every rule is derived from some objective construction involving sets from the universe, but not the universe itself. Such constructions must be *objective*, in the sense that they are completely determined independently of the universe. On the other hand, an objective construction may involve a transcendental or creative element that goes beyond the given sets, and this is certainly the case in the construction of the set ω referred to above. The basic premise in this work is that in spite of such creativeness, a construction involving a reference to the universe as a complete totality is not objective, and as a consequence is not admissible.

A non-objective construction depending on the universe takes place whenever a condition is used that contains non-local quantifiers. This may happen, for example, in applications of the separation axiom. Another form of reference to the universe occurs in abstractions where the range of the variable is not bounded. This is the case with the standard definition of the power-set.

The predicativity principle implies that non-local quantification over the universe is not a legitimate construction. This restriction is strictly enforced in this work, by imposing local quantification in the rules of the system. As a consequence, several classical proofs are not available, in particular definitions by transfinite recursion. *In fact, induction and recursion are taken as primitive rules.*

The restriction to local quantification is required in the applications of the rules to generate new predicates and sets. In the general language of the theory, non-local quantification is allowed since we are dealing with assertions about arbitrary sets in the universe. Still, it is required that in dealing with non-local expressions, deductions be restricted to intuitionistic logic.

The formal system is organized as a collection of rules that can be used to introduce set operations and set predicates. In particular, the rule of set induction introduces set predicates, and the rule of set recursion introduces set operations relative to predicates previously defined by induction. Each rule is derived from an objective construction in the informal system, where we essentially assume the absolute model. From the informal construction, formal axioms are derived that become elements of the formal rule. These axioms constitute the only connection between the informal and the formal theories. In the formal system only the formal axioms are available.

Although the complete system is not equivalent to Zermelo–Fraenkel, it still represents a good approximation that can be used in many applications. The main advantage is that it is derived from predicative constructions which provide a guarantee of internal consistency. In fact, each construction in the system is obtained by a sequence of applications of the primitive rules, which are derived by logical-mathematical constructions in the absolute model.

The need for a predicative critique of set theory has been expressed many times, in some cases in the framework of an intuitionistic approach to mathematics. *I am not committed to any particular philosophy or ideology, and this work is intended only as a pragmatic elaboration of objective constructions.* Weyl has given a very forceful expression of this position by pointing to the roots of the paradoxes in the faith in the universe of sets "as a *closed* aggregate of objects existing in themselves" (my italics).

The implementation of this predicative program is very much dependent on the operational framework in which the theory is formulated. For example, in standard set theory the rule of substitution is actually derived by non-local existential quantification (more precisely, substitutions are eliminated via non-local existential expressions). Since we take set operations and substitution as primitive, there is no need to invoke any type of elimination.

Still, the operational framework is an essential part of the system proposed here, and not just a syntactical device necessary to implement the predicative program. There are good theoretical reasons for approaching set theory as a system of constructions. Furthermore, there are practical reasons, for the usual formalizations via existential axioms are difficult to handle and, in practice, obscure the constructive contents implicit in the axioms. I expect that an operational approach to set theory will be welcomed by practitioners more interested in applications to standard mathematical theories.

A delicate point in this discussion is the determination of the proper logic that can be used in set theory. A careful formalization is advanced in chapters 9 and 10, where I propose that in dealing with global expressions we must use intuitionistic logic, while in dealing with local expressions we can rely on classical logic. A convenient arrangement is organized via the well-known Gentzen rules.

I address this work to those mathematicians and philosophers who are concerned about the role of set theory in the foundations of mathematics. The set theory specialist may also find new ideas of some interest, particularly the formulation of very general rules of set induction and set recursion.

I thank Syracuse University for the support of this project. In particular, Vera Elaine Weinman has provided generous technical advice and has supervised the general preparation of the manuscript well beyond the requirements of her position as editorial assistant in the School of Computer and Information Science. Special thanks to an anonymous reviewer for very helpful suggestions and criticisms.

My family has provided all the patience and understanding that my concentration on this work has required. I am very grateful to all of them.

Chapter 1

Operations and Predicates

We study sets by means of a first order system involving set operations, set predicates, and the usual first order constructions: propositional connectives and quantifiers. We assume the reader has some familiarity with the symbolism of mathematical logic.

We assume a universal domain D where the elements are sets. We consider D an ambiguous totality, whose elements are not completely determined. If the reader has difficulties with this idea, there is no need to worry about it. It only means that some constructions involving quantification over D are not admissible in our system.

The purpose of the theory is to introduce meaningful constructions that can be applied to arbitrary sets in D to generate new sets that are also elements of D. Such constructions are characterized by specific axioms, and by using them we can prove general properties of the sets in D. So, D is an ever present reality in $\mathbb{G}$, and the set variables in our language are intended to denote arbitrary elements of D.

Although D is an ambiguous totality, we assume that each set in D is *absolute* in the sense that a set is a complete totality, independent of the domain D. So, set theory is the study of objects that are complete totalities in a context (the universal domain D) that it is not a complete totality.

This restriction is enforced in different ways, in particular by not allowing set constructions involving global quantification over the universe.

1.1 The System $\mathbb{G}$

We call $\mathbb{G}$ (for Georg Cantor, the founder of set theory) the system of predicative set theory we study in this work This is a first order system in the sense of mathematical logic, and includes predicates, operations, propositional connectives, and quantifiers. We assume the reader has some familiarity with this terminology. In this connection we use the following notation: $\neg$ (*negation*), $\vee$ (*disjunction*), $\wedge$ (*conjunction*), $\rightarrow$ (*conditional*), $(\forall X)$ (*universal global quantifiers*), $(\exists X)$ (*existential global quantifiers*), $(\forall X \in Y)$ (*universal local quantifier*), $(\exists X \in Y)$ (*existential local quantifier*). We use also $\equiv$ (*biconditional*), which is assumed to be defined in the usual way using conjunction and conditional.

This formalism is used in several ways. First, we can write expressions (well-formed formulas) referring to the sets in the universe, and the universe itself (by means of global formulas, involving global quantifiers). Some of these expressions are axioms of the theory, and from the axioms we can deduce theorems. Second, the formalism is used in the definition of set operations and set predicates.

Remark 1.1 A complete formalization of the system is provided in Chapter 9, where the following categories are defined by simultaneous syntactical induction: terms, predicates, operations, local formulas, global formulas, and axioms. □

Remark 1.2 The sets in the universe are assumed to be complete totalities. In fact, this is the essence of set theory and the basis of its success. There has always existed a sector of the mathematical community (mainly in intuitionistic mathematics) that rejects this assumption for infinite sets, for a number of reasons. They do not mean that it could cause inconsistencies, rather that infinite sets are mental constructions that do not support a claim of totality. In this work this position is rejected, and we show that the set ω of natural numbers is derived by an absolute construction (see Section 7.1). Still, constructions going beyond the natural numbers require new principles that are discussed in Chapter 8 and the Appendix. □

Some remarks concerning the role of logic in set theory are in order. The current practice is to use set theory to provide the foundations of classical and intuitionistic logic. This means that the validity of logical deduction is determined via set interpretations, and the validity of the logic comes from the completeness relative to such interpretations.

In principle, we take the position that the logic of set theory is an integral part of the theory and both must be defined simultaneously. Concerning the system $\mathbb{G}$ we think that a distinction must be introduced between global expressions and local expressions. For local expressions it seems that classical logic is adequate. In the case of global expressions it appears that some principles are not consistent with the fundamental assumptions about the universe D, and we are inclined to rely on intuitionistic logic.

This matter is discussed in Chapters 9 and 10, where both logics are formalized via the well-known Gentzen rules. In Corollary 10.4.1 we prove a classical reduction, in the sense that local theorems proved in classical logic can also be proved under intuitionistic logic. Most theorems proved in this book are local formulas and some of them are global formulas of the form $((\forall Z)\phi(Z) \rightarrow (\forall Z)\psi(Z)$ (the most important example being Theorem 3.3). In every case these results are proved using straight classical logic and the reduction theorem in Chapter 10 shows how to obtain a valid intuitionistic proof in the sense defined there, even for global formulas (see Example 10.3). This means that, as far as the set theoretical results proved in this book, the reader may ignore entirely the distinction between classical logic and intuitionistic logic. We use only classical logic, which is the standard logic in mathematics, with the understanding that a reduction to intuitionistic logic is always available. Note that the system of intuitionistic logic in Chapter 10 allows for classical axioms on local formulas (see rule **ST 15** in Section 9.1).

We shall use as variables ranging over the domain D (*set variables*), the symbols: $A, B, \ldots, Y, Z$ (italic capital letters). Besides the notion of set (implicit in the use of the variables) there are two primitive notions in the system $\mathbb{G}$: the binary membership relation $\in$, written $X \in Y$, and the equality relation $=$, written $X = Y$. Equality has the usual properties in mathematical systems. If $X \in Y$ holds, we say that X is an *element of* Y. These two primitives are related by the following principle of extensionality, which is the first axiom of the system $\mathbb{G}$.

Principle of Extensionality If X and Y are sets with the same elements, then $X = Y$.

Remark 1.3 The principle of extensionality can be expressed using the logical formalism,

$$(\forall V \in X)V \in Y \wedge (\forall V \in Y)V \in X \rightarrow X = Y.$$

This is actually a local formula, for it does not involve quantification over the universe. As an axiom it is assumed to be valid over the universe D, in the sense that the axiom is true for arbitrary sets X and Y in the universe.

All axioms in the system $\mathbb{G}$ are written as local expressions (not involving global quantifiers) with free variables ranging over the universe. Each axiom is intended to be true for all possible values of the free variables, so the real axioms are the universal closure of the local expressions. This shows that the universe D is an essential element of the system $\mathbb{G}$, although this does not imply that it is a complete totality. □

1.2 Basic Rules

The system $\mathbb{G}$ is organized as a system of rules, and each rule introduces set operations or set predicates (the set induction rule in Chapter 3 introduces a set predicate and a set operation). The system is open in the sense that it can be expanded with new rules. Each rule is preceded by an informal primitive construction and contains a number of formal axioms that are derived from the primitive construction.

Remark 1.4 Primitive constructions must satisfy a number of informal requirements. First, a primitive construction must be consistent with the principle of extensionality, so the result of the construction must be determined by the inputs and their elements. Second, it must be objective, although this does not exclude some transcendental completion. For example, the set ω is generated by the fundamental sequence $0, 1, 2, ...$, but requires a final action to put together all the components of the sequence. In particular, the objectivity condition excludes any reference to the universe as a totality, since there is no objective construction of the universe. □

A *k-ary set operation*, where $k \geq 0$, is an operation that can be applied to k sets $X_1, ..., X_k$ to produce a new set Y. The application is written

$$f(X_1, ..., X_k) = Y.$$

Remark 1.5 Note that set operations are total operations in the sense that the application $f(X_1, ..., X_k)$ is assumed to be defined and denotes a set for any value (in the universe) of the variables $X_1, ..., X_k$. In fact, allowing for partially defined set operations would create serious problems in the theory, for the expression of the condition "$f(X_1, ..., X_k)$ is defined" requires an existential quantifier ranging over the universe (i.e., $(\exists Y) f(X_1, ..., X_k) = Y$), and the condition: "$f(X_1, ..., X_k)$ is undefined" is still more involved. In some situations we may want to have partially defined operations, in which case we must use functions, to be defined in the next chapter. □

In several cases the formal axiom for a rule that introduces the operation f will have the form of an *abstraction* where f is described extensionally as

$$Z \in f(X_1, ..., X_k) \equiv ...Z...X_1...X_k....$$

In this expression there is an implicit non-local universal quantifier on the variable Z, and this appears to contradict our assumptions on the universe. In fact, we do not use abstractions as definitions. They are only axioms that describe general properties of the sets introduced by the rules. In this work all the abstractions come from the replacement rule in Chapter 2 and are described in the standard brace formalism (see Remark 2.2).

A *k-ary set predicate p*, $k \geq 0$ is also an operation that is applied to k sets $X_1, ..., X_k$, and the result of the application is either *true* or *false*. The application is written in the form

$$p(X_1, ..., X_k),$$

and this expression is asserted when the value is *true* and negated when the value is *false*. As usual, set predicates are assumed to be total, so $p(X_1, ..., X_k)$ is defined for any values of $X_1, ..., X_k$.

Set predicates are also introduced by rules that in general involve a number of parameters and previous constructions, on the basis of primitive constructions. From the primitive constructions we derive formal axioms. In some cases we assume that the axioms are actually identical with the primitive constructions.

Remark 1.6 Very often a sequence of variables $X_1, ..., X_k$ will be contracted to a boldface symbol $\mathbf{X}$. So the application of the k-ary operation f can be written $f(\mathbf{X})$, and the application of a k-ary predicate p can be written $p(\mathbf{X})$. The case $k = 0$ is important when we consider a 0-ary set operation f, which can be identified with a set A such that $f(\ ) = A$. In this case we can write $f(\mathbf{X}) = A$, assuming that $\mathbf{X}$ is an empty sequence of variables. □

We call the rules of the system $\mathbb{G}$ *basic rules*, and the operations and predicates introduced by such rules *basic operations* and *basic predicates.* Most of them are familiar from standard set theory, where they are usually derived from strong impredicative axioms. Here they are taken as primitive and assumed to be consistent with our predicative approach. Later (see Section 2.5) we identify a special group of basic rules that we call elementary rules, and also the elementary operations and predicates that are generated by such rules.

Besides basic operations and predicates, we have *general set operations* and *general set predicates*, which are derived from basic set operations and predicates by substitution with arbitrary sets. These constructions are an essential part of the theory, as they provide the machinery to connect the formal system with the universe. For example, if f is a binary basic set operation, and A is a set (not necessarily basic) we can introduce a unary general set operation g in the form

$$g(X) = f(A, X).$$

The set A may be known from some discussion or proof. The general operation g is not basic, but still is a legitimate operation. Similarly, from a basic predicate p we can derive by substitution new general predicates, which may play a crucial role in proofs by induction.

Some of the basic rules are derivation rules that introduce new set operations and set predicates from given set operations and predicates. By definition, *the operation or predicate introduced by a basic rule is basic whenever the given operations and predicates are basic.* On the other hand, the basic rules are meaningful also in case the given operations and predicates are general (in fact, they are meaningful for arbitrary operations and predicates). It can be shown that in every such case the operation or predicate introduced by the rule is also general. This is obvious in most rules, and it follows from Theorem 1.1. The rules of set induction and set recursion require some special argument (see Remarks 3.1 and 5.1).

With this understanding we write the rules assuming given general operations and predicates. If they are basic, then by definition it follows that the operation or predicate introduced by the rule is basic. If some of them are general, then we conclude that the operation or predicate introduced by the rule is general in the sense of the definition given above in terms of substitution with fixed sets.

In a number of places we discuss an arbitrary set operation or an arbitrary set predicate. These operations and predicates are always intended to be general in the sense explained above.

To help the reader to have a picture of the whole system we list here all the primitive rules of the system $\mathbb{G}$.

The empty set rule (Chapter 1).

The equality rule (Chapter 1).

The membership rule (Chapter 1).

The insertion rule (Chapter 1).

The projection rule (Chapter 1).

The operation substitution rule (Chapter 1)

The predicate substitution rule (Chapter 1).

The boolean connectives rule (Chapter 1).

The universal local quantification rule (Chapter 1).

The existential local quantification rule (Chapter 1).

The replacement rule (Chapter 2).

The set induction rule (Chapter 3).

The set recursion rule (Chapter 5).

The omega rule (Chapter 7).

The ordinal permutation rule (Chapter 8).

The enumeration rule (Appendix).

Note that the enumeration rule can be derived from the ordinal permutation rule, but in the Appendix it is proposed as an independent rule. Many of these rules correspond to very simple syntactical constructions in standard set theory (projection, substitution, connectives, quantifiers). The crucial rules are: replacement, induction, recursion, omega and ordinal permutation.

1.3 Initial Rules

Some of the rules of the system $\mathbb{G}$ are *initial rules*, because they introduce definite set operations and set predicates. Furthermore, there are *derivation rules*, where new set operations or set predicates are derived from given set operations and predicates. All rules in this section are initial rules. Two more initial rules, the omega rule and the ordinal permutation rule, are introduced later (see Chapters 7 and 8). All the other rules are derivation rules. There are also derived rules (definition by cases, local abstraction and bounded minimalization) that are not a part of the basic system (see Sections 2.4 and 6.5).

The Empty Set Rule We introduce a 0-ary set operation, which we identify with a set (*the empty set*) denoted by $\emptyset$. We assume this set is given by a primitive construction that determines that $\emptyset$ has no element. It is characterized by the following axiom.

The Empty Set Axiom: $X \notin \emptyset$.

Remark 1.7 As explained above, the set $\emptyset$ is introduced by a primitive construction where it is determined that $\emptyset$ has no element. This condition refers to $\emptyset$ itself, not to the universe. It follows from the condition that if X is any set in the universe, then $X \notin \emptyset$, and this is what the formal axiom asserts. In fact, it follows that if X is an arbitrary object, not necessarily in the universe, then $X \notin \emptyset$. □

We introduce also two binary initial predicates, corresponding to the equality relation and the membership relation. In these two predicates we do not invoke explicitly any primitive construction, since we take the notion of set, membership, and equality as primitive in our system. In this context the formal axioms can be considered as "definitions" of the predicates.

The Equality Rule We introduce a binary set predicate **EQ** that satisfies the following axiom.

The Equality Axiom:

$$\mathbf{EQ}(X, Y) \equiv X = Y.$$

The Membership Rule We introduce a binary set predicate **EP** which satisfies the following axiom.

The Membership Axiom:

$$\mathbf{EP}(X, Y) \equiv X \in Y.$$

Remark 1.8 Although "**EQ**" and "**EP**" are the official names of the predicates equality and membership, we shall continue to use the usual symbols "$=$" and "$\in$". □

Next, we introduce an initial rule for a binary set operation **ins** (*the insertion operation*). The primitive construction for this operation determines that, given sets X and Y, then $\mathbf{ins}(X, Y)$ is the set obtained by inserting Y as a new element in the set X, with the understanding that if $Y \in X$, then the result is X again. This is a crucial construction where X must be a set, while Y is in fact an arbitrary object. It means that any set X can be changed to another set that contains any given set Y. We proceed now to write the formal rule with the corresponding axiom. As usual the axiom is not intended as a "definition" of the operation (see Remarks 1.3 and 1.7).

The Insertion Rule We introduce a binary set operation denoted by **ins**, although in practice we shall use the in-fixed notation $\mathbf{ins}(X, Y) = X; Y$. This operation satisfies the following axiom.

The Insertion Axiom:

$$Z \in X; Y \equiv Z \in X \vee Z = Y.$$

Finally, we introduce an infinite number of trivial projection operations that are necessary for technical reasons (see Example 1.3).

The Projection Rule For any $k \geq 1$ and i such that $1 \leq i \leq k$, we introduce a k-ary projection operation $\mathbf{pro}_i^k$, which satisfies the following axiom.

The Projection Axiom:

$$\mathbf{pro}_i^k(X_1, ..., X_k) = X_i.$$

EXERCISE 1.1 Prove that $\emptyset \neq \emptyset; \emptyset$.

EXAMPLE 1.1 Both the empty set and insertion operations are basic by definition. If we set $f(Y) = \emptyset; Y$ we get a new general operation. Later, we extend the rules and f will become a basic operation, but at this stage we can only say that f is a general operation. Clearly, $f(Y)$ denotes a set with only one element, namely, Y. If we set $g(Y) = \emptyset; f(Y)$, then g is not even a general operation (but later it will become a basic operation). Still, g is a well-defined operation in the sense that, for any given set Y, $g(Y)$ denotes a well-defined set. □

EXERCISE 1.2 Let f be the operation in Example 1.1. Is it possible to have a set Y such that $Y = f(Y)$?

1.4 Substitution

There are two substitution rules, one for set operations and the other for set predicates. In both cases we assume that several operations and predicates are given, and the rule introduces a new operation or predicate. If the given operations and predicates are basic, it follows that the new operation or predicate is also basic. If the given operations and predicates are general, rather than basic, the application of the rule is still meaningful, but no claim of being basic is made. Still, from Theorem 1.1 it follows that the operation or predicate is general.

The Operation Substitution Rule Assume an m-ary general set operation f is given, $m \geq 0$, and also m k-ary general set operations $g_1, ..., g_m$, $k \geq 0$, are given. We introduce a new k-ary set operation h that satisfies the following axiom.

The Operation Substitution Axiom:

$$h(\mathbf{X}) = f(g_1(\mathbf{X}), ..., g_m(\mathbf{X})).$$

We say that h is obtained from $f, g_1, ..., g_m$ by *operation substitution.*

The Predicate Substitution Rule Assume an m-ary general set predicate q is given, $m \geq 0$, and also that m k-ary general set operations $g_1, ..., g_m$, $k \geq 0$ are given. We introduce a new k-ary set predicate p that satisfies the following axiom.

The Predicate Substitution Axiom:

$$p(\mathbf{X}) \equiv q(g_1(\mathbf{X}), ..., g_m(\mathbf{X})).$$

We say that p is obtained from $q, g_1, ..., g_m$ by *predicate substitution.*

Remark 1.9 It is not necessary to discuss the primitive constructions supporting the substitution rules. Since we are dealing with total operations and predicates, the substitution expressions are completely transparent and, in fact, we may consider that the axioms are actually definitions of the constructions. □

EXAMPLE 1.2 We can use operation substitution with $\emptyset$, which means $m = 0$ in the rule. Hence for any k we can introduce a basic k-ary set operation f such that

$$f(\mathbf{X}) = \emptyset.$$

Instead of $\emptyset$ we can use an arbitrary general set A. In any case, the operation f is general. If A is basic, then f is basic (see Theorem 1.1). □

EXAMPLE 1.3 Using predicate substitution in the predicate **EP** we can introduce a basic k-ary predicate p such that

$$p(\mathbf{X}) \equiv \mathbf{EP}(\mathbf{pro}_i^k(\mathbf{X}), \mathbf{pro}_j^k(\mathbf{X})),$$

where $1 \leq i \leq k$ and $1 \leq j \leq k$. It follows that

$$p(\mathbf{X}) \equiv X_i \in X_j.$$

Instead of **EP** we can use an arbitrary general binary predicate q. In any case, the predicate p is general. If q is basic, then p is basic. □

Remark 1.10 The rule of predicate substitution is completely natural in the system $\mathbb{G}$, since operations are primitive constructions in the system. In standard set theory, where operations usually are not primitive, such substitutions are expressed by means of global existential quantifiers, which are not available in the system $\mathbb{G}$. In some applications the rule of predicate substitution is crucial (see Theorem 2.6). □

1.5 Logical Rules

We need logical rules to generate new set predicates, for the rule of predicate substitution is clearly insufficient to go much further than the initial predicates of Section 1.3. We include here the boolean connectives and the local quantifiers. The boolean connectives are introduced in a fairly informal way, since they are determined by well-known truth tables, and there is no need to enter into the details here. We are a bit more formal concerning the quantifiers, as they are used only in a local version.

The Boolean Connectives Rule We can obtain new set predicates from given predicates by using the propositional connectives described in Section 1.1, which we assume are determined by the corresponding truth tables. For example, negation is an operation such that, given a k-ary set predicate q, we introduce a k-ary set predicate p with the formal axiom,

$$p(\mathbf{X}) \equiv \neg q(\mathbf{X}).$$

Similarly, disjunction is an operation such that, given k-ary set predicates q_1 and q_2, we introduce a new k-ary set predicate p with the formal axiom:

$$p(\mathbf{X}) \equiv q_1(\mathbf{X}) \vee q_2(\mathbf{X}).$$

The connectives conjunction and conditional are treated in the same way. Recall that the biconditional is assumed to be defined from conjunction and conditional.

We can derive set predicates by using quantifiers, but only in a local format. This means that the range of the quantifier must be a given set. This is consistent with the assumptions about the universe, since non-local quantification would imply that the universe is considered a complete totality. Still, global quantification is available in the language of the theory and, in fact, we can prove theorems that are expressed by global expressions (see Theorem 3.3). In any case, only local quantification is available in the definition of set predicates, and this affects the definition of set operations, since they are interdependent.

The Universal Local Quantification Rule If q is a given $(k+2)$-ary set predicate we can introduce a new $(k+1)$-ary set predicate p that satisfies the following axiom.

The Universal Local Quantification Axiom:

$$p(Z, \mathbf{X}) \equiv (\forall Y \in Z) q(Y, Z, \mathbf{X}).$$

We say that p is obtained from q by *universal local quantification.* As usual, if $Z = \emptyset$, then $p(Z, \mathbf{X})$ is true.

The Existential Local Quantification Rule If q is a given $(k+2)$-ary set predicate we can introduce a new $(k+1)$-ary set predicate p that satisfies the following axiom.

The Existential Local Quantification Axiom:

$$p(Z, \mathbf{X}) \equiv (\exists Y \in Z) q(Y, Z, \mathbf{X}).$$

We say that p is obtained from q by *existential local quantification.* As usual, if $Z = \emptyset$, then $p(Z, \mathbf{X})$ is false.

EXAMPLE 1.4 Both the substitution rules and the logical rules are introduced in a rigid format that creates some problem in applications. For example, the introduction of a predicate p in the form,

$$p(X, Y) \equiv (\exists Z \in X) \mathbf{EP}(Y, Z),$$

is not consistent with the rule of existential quantification, so the predicate p is not derived from **EP** by existential local quantification. On the other hand, we can use predicate substitution with the predicate **EP** to define a new basic predicate **EP**′ such that

$$\mathbf{EP}'(Z,X,Y) \equiv \mathbf{EP}(\mathbf{pro}_3^3(Z,X,Y), \mathbf{pro}_1^3(Z,X,Y)).$$

Now we can obtain p from **EP**′ using existential local quantification in the form,

$$p(X,Y) \equiv (\exists Z \in X)\mathbf{EP}'(Z,X,Y),$$

and this shows that p is a basic predicate. □

Obviously, we do not want to have to go in every case through the analysis performed in the preceding example. In fact, we can avoid these complications by using a well-known technique whereby several applications of the rules can be described by general expressions.

EXERCISE 1.3 Use initial predicates, projection, substitution, and logical rules to introduce a basic binary set predicate p that satisfies the relation $p(X,Y) \equiv X \notin X \wedge X \in Y$ for all sets X and Y.

1.6 Basic Terms

The preceding operations, predicates, and rules are only fragments of the whole theory we want to present in this work. Before defining more rules it is convenient to introduce a general technique where the applications of several rules are expressed by terms involving variables and previously introduced constructions. These expressions we call *basic terms*, and are defined inductively by a number of formal rules. There are two types of basic terms, *set terms* and *boolean terms*. The following inductive rules define both types of terms.

BT 1: A set variable is a basic set term.

BT 2: If f is a k-ary operation and $U_1, \ldots, U_k$ are basic set terms, then $f(U_1, \ldots, U_k)$ is a basic set term.

BT 3: If p is a k-ary predicate and $U_1, \ldots, U_k$ are basic set terms, then $p(U_1, \ldots, U_k)$ is a basic boolean term.

BT 4: If U and V are basic boolean terms, then $\neg U$, $(U \vee V)$, $(U \wedge V)$, $(U \rightarrow V)$, and $(U \equiv V)$ are basic boolean terms.

BT 5: If U is a basic boolean term, V is a basic set term, and Y is a variable that does not occur free in V, then $(\forall Y \in V)U$ and $(\exists Y \in V)U$ are basic boolean terms.

As usual, the variable X *is bound* in a construction by rule **BT 5** or, more precisely, the occurrences of Y in the scope of the quantifiers are bound. An occurrence of a variable that is not bound is said to be *free*. Note that no variable in the term V in rule **BT 5** is bound by the construction, and in fact, all variables in V remain free when the rule is applied. Still, variables in V may become bound by a later quantifier.

The semantics for a basic terms U is self-explanatory and can be derived from the notation. It depends essentially on the values of the free variables in U, on the denotation of the symbol f in rule **BT 2**, and on the denotation of the symbol p in rule **BT 3**. In principle, f is an arbitrary set operation and p is an arbitrary set predicate. We are interested in the semantics when f is either a basic or a general operation, and p is a basic or a general predicate.

Let **X** be a list containing all the free variables in U (and possibly variables that do not occur free in U). The meaning of U is determined by the denotations of the variables in **X** which, of course, we assume are sets. If U is a basic set term, then the denotation of U is a set, and if U is a basic boolean term, then the denotation is a truth value. These properties can be proved easily by induction on the construction of the term U.

The preceding discussion shows that we can use basic terms to define set operations and set predicates. In fact, if U is a set term and **X** is a list of variables of length k that contains all the free variables in U, then U induces a k-ary set operation f_{U} such that

$$f_{\mathrm{U}}(\mathbf{X}) = \mathrm{U},$$

and if U is a boolean term, then U induces a k-ary predicate p_{U} such that

$$p_{\mathrm{U}}(\mathbf{X}) \equiv \mathrm{U}.$$

Theorem 1.1 *Let* U *be a basic term and* $\mathbf{X}$ *a list of variables containing all the variables occurring free in* U. *Then:*

(i) *If* U *is a set term and all the set operations and set predicates in* U *are basic (general), then the set operation* f_{U} *induced by* U *is basic (general).*

(ii) *If* U *is a boolean term and all the set operations and set predicates in* U *are basic (general), then the predicate* p_{U} *induced by* U *is basic (general).*

PROOF. First we assume that all the operations and predicates in U are basic, and prove the theorem by induction on the construction of the term U. Rule **BT 1** follows using the projection rule, for if U is the variable X_i in the list $\mathbf{X}$, then $f_{\mathrm{U}}(\mathbf{X}) = \mathbf{pro}_i^k(\mathbf{X})$. Rule **BT 2**, **BT 3**, and **BT 4** follow immediately using the induction hypothesis. In rule **BT 5** assume U is the term $(\forall Y \in \mathrm{V})\mathrm{U}_1$. By the induction hypothesis $q'(Y, Z, \mathbf{X}) \equiv \mathrm{U}_1$ and $f(\mathbf{X}) = \mathrm{V}$ are basic (where Z is a new variable not occurring in U_1). From q' we get $q(Z, \mathbf{X}) \equiv (\forall Y \in Z)q'(Y, Z, \mathbf{X})$ and finally $p_{\mathrm{U}}(\mathbf{X}) \equiv q(f(\mathbf{X}), \mathbf{X})$ (this substitution can be obtained using the predicate substitution rule and the projection rule, see Example 1.3).

If U is a set term, and some of the operations and predicates in U are general but not basic, we eliminate such occurrences in the following way. First, each one of the non-basic operations or predicates comes from basic operations and predicates by substitutions with fixed sets. We write all the required sets in a list: $V_1, ..., V_s$. Second, we take s new variables $Y_1, ..., Y_s$, and replace every general operation and predicate by its substitution definition, using the variable Y_i whenever the set V_i is required in the substitution, $i = 1, ..., s$. In this way we obtain a new basic term U′ where all the free variables are in the list $\mathbf{X}, Y_1, ..., Y_s$ and all the set operations and set predicates are basic, so by the first part of the theorem the set operation or set predicate induced by U′ is basic. Finally, we have $f_{\mathrm{U}}(\mathbf{X}) = f_{\mathrm{U}'}(\mathbf{X}, V_1, ..., V_s)$. We use similar substitution if U is a boolean term. □

Remark 1.11 Note that the basic terms defined by rules **BT 1** to **BT 5** correspond to the basic rules introduced up to this point. New terms are defined later, corresponding to some of the new basic rules, and Theorem 1.1 will be valid for the extension. Still, the set induction and set recursion rules cannot be replaced with basic terms. □

We apply Theorem 1.1 to define several new set operations and set predicates. We assume $k \geq 1$.

1.6.1 $\{X\} = \emptyset; X$ (*singleton operation*).

1.6.2 $\{X_1, ..., X_k, X_{k+1}\} = \{X_1, ..., X_k\}; X_{k+1}$ (*general insertion operation*).

1.6.3 $< X, Y >= \{\{X\}, \{X, Y\}\}$ (*the ordered pair operation*).

1.6.4 $S(X) = X; X$ (*the successor operation*).

1.6.5 $Y \subseteq X \equiv (\forall Z \in Y) Z \in X$ (*Y is a subset of X*).

1.6.6 $Y \subset X \equiv Y \subseteq X \wedge Y \neq X$ (*Y is a proper subset of X*).

1.6.7 $\mathbf{TR}(Z) \equiv (\forall Y \in Z) Y \subseteq Z$ (*Z is transitive*).

1.6.8 $\mathbf{DR}(Z) \equiv (\forall Y \in Z)(\forall X \in Z)(\exists V \in Z)(Y \subseteq V \wedge X \subseteq V)$ (*X is directed*).

EXERCISE 1.4 Let Z and X be sets. Prove the following conditions are equivalent:

(a) $Z \in \{X\}$

(b) $Z = X$

EXERCISE 1.5 Assume $\{X\} = \{Y\}$. Prove $X = Y$.

EXERCISE 1.6 Prove that the sets $\emptyset$ and $\{\emptyset\}$ are transitive. Is the set $\{\{\emptyset\}\}$ transitive?

EXERCISE 1.7 Let X and Y be sets. Prove the following conditions are equivalent:

(a) $Y \in S(X)$

(b) $Y \in X \vee Y = X$

EXERCISE 1.8 Assume the set X is transitive. Prove that $S(X)$ is also transitive.

EXERCISE 1.9 Prove that $<X, X> = \{\{X\}\}$.

EXERCISE 1.10 Let X and Y be sets where $Y \subseteq X \subseteq S(Y)$ holds. Prove that either $X = Y$ or $X = S(Y)$.

EXERCISE 1.11 Let Y be a transitive set and X an arbitrary set. Prove that the following conditions are equivalent:

(a) $S(X) \subseteq Y$.

(b) $X \in Y$.

EXERCISE 1.12 Prove that if X is transitive and directed, then $S(X)$ is also directed.

EXERCISE 1.13 Let X, Y be sets. Prove the following conditions are equivalent:

(a) $\{X, Y\}$ is directed.

(b) $X \subseteq Y \vee Y \subseteq X$.

1.7 Notes

Cantor's position concerning the nature of the universe of sets cannot be determined. Given his initial naive disposition to take as sets any extension of a property, it seems that he was inclined to consider the universe as a complete totality. On the other hand, his reaction to the Burali-Forti paradox in Cantor (1895), where he restricted the class of all ordinals to the second class, suggests an evolution toward a more conservative position. For a complete discussion of Cantor's ideas, see Cavaillès (1962) and Hallett (1984).

Our personal position (that Moschovakis (1991) refers to as a large view of the universe) is one of several predicative approaches available in the literature (see Myhill (1975)). The opposite impredicative position, where the universe is considered a complete totality, is popular among mathematicians of a classical persuasion. It appears to us that many of them are aware of the logical difficulties the impredicative position implies, but still do not find feasible to impose reasonable restrictions that preserve the integrity of traditional set theory. We try in this work to reach a balance via the ordinal permutation rule in Chapter 8. Still, this rule is only a compromise where the universe of ordinals is invoked, although only in a potential way. On the other hand, the restriction to local quantification has been explored in a number of places, particularly in Barwise (1975).

The universal domain is a complete totality in standard set theory that allows non-local quantification over the universe. This being the case, it is not surprising that attempts have been made to give an objective determination of the universe. The widely accepted one comes from Shoenfield (1967) and involves a transfinite iteration where at each level a new set is generated, which is the power-set of the set generated at the preceding level (see Definition 8.3.2 in Chapter 8 and Boolos (1971)). This is essentially a predicate (equivalent to the predicate **WF** we define by induction in Chapter 3), and the extension of a predicate is a reflection of the universe. For this reason we do not think that this construction provides an objective determination of the universe.

The impredicative conception of the universe is strongly objected in Weyl (1949), as involving the assumption of a closed collection derived only from constructive possibilities.

On the other hand, there have been interesting theories where, rather than a unique universe, a hierarchy of universes is postulated with each universe a set of the universes in a higher level of the hierarchy (Sochor (1984)).

By means of the logical rules in this chapter we generate structures that correspond to the well-formed formulas in standard set theory, with the important restriction that we allow only local quantification. On the other hand, in standard set theory there is no need for special rules of substitution, since substitution is actually built into the rules defining well-formed formulas. We reach the same level of convenience via the introduction of terms. So both presentations are essentially equivalent, if we ignore the restriction to local quantification.

The restriction to local quantification is critical, but the rules of induction and recursion (see Chapters 3 and 5) provide for extra machinery which is also available in the standard formalism (see Section A.4).

Chapter 2

Replacement

In this chapter we present one of the fundamental constructions in the theory of sets, where new set operations are derived from given set operations and predicates. In standard set theory this construction is usually expressed as an axiom of replacement. For us it is a derivation rule supported by a primitive construction, which we claim is objective and independent of the universe.

The rule we introduce below goes beyond the usual interpretation of replacement and includes a version of the well-known union axiom.

2.1 The Replacement Rule

We start with a discussion of a primitive construction in a very simple situation where a unary set operation h is given, and we describe a set operation $f(X)$ by the following process. Given an input set X and some set $Y \in X$, we apply h to Y and generate all the elements in the set $h(Y)$. We collect all the elements obtained in this way for every $Y \in X$, and form a set X'. Finally, we set $f(X) = X'$.

We use the following equivalence to describe the elements of the set X':

$$Z \in X' \equiv (\exists Y \in X) Z \in h(Y).$$

The set X' is the result of an objective process completely determined by the elements of the set X and the operation h. If we assume that h is independent of the universe, it follows that the operation f is given by an objective primitive construction independent of the universe, so it is a legitimate rule in the theory of sets.

The preceding situation can be generalized by assuming sets $X_1, \ldots, X_s$, and an s-ary operation h. Now we generate a set X' that is given by the following equivalence:

$$Z \in X' \equiv (\exists Y_1 \in X_1) \ldots (\exists Y_s \in X_s) Z \in h(Y_1, \ldots, Y_s).$$

A more general construction can be obtained by setting X' to be the set whose elements are in the set $h(Y_1, \ldots, Y_s, \mathbf{X})$, and there are given set operations $g_1, \ldots, g_s$ such that $Y_1 \in g_1(\mathbf{X}), \ldots, Y_s \in g_s(\mathbf{X})$.

A final generalization is obtained by introducing a condition p which is imposed on the elements $Z, Y_1, \ldots, Y_s, \mathbf{X}$.

The Replacement Rule Let $g_1, ..., g_s$ be k-ary set operations, $s \geq 0$, $k \geq 0$, h an $(s+k)$-ary set operation, and p an $(s+k+1)$-ary set predicate. We introduce a k-ary set operation f that satisfies the following axiom.

The Replacement Axiom:

$$Z \in f(\mathbf{X}) \equiv (\exists Y_1 \in g_1(\mathbf{X}))...(\exists Y_s \in g_s(\mathbf{X}))(Z \in h(\mathbf{Y}, \mathbf{X}) \land p(Z, \mathbf{Y}, \mathbf{X}))$$

where $\mathbf{Y} = Y_1, ..., Y_s$. Note that the replacement axiom in this rule is different from the standard replacement axiom that occurs in standard set theory. Later we give a short argument to show that the former can be derived from the latter (see Remark 2.3).

The rule can be applied even if some of the given operations or predicates are missing. For example, if $s = 0$ there are no operations $g_1, ..., g_s$. If the predicate p is missing we assume it is a true predicate that is always satisfied.

What is essential is the operation h, for otherwise the variable Z becomes global.

The application of this rule as given may be rather cumbersome, since it requires the identification of several set operations and one set predicate. We can avoid these complications by using set terms $\mathrm{W}_1, ..., \mathrm{W}_s$ in place of $g_1(\mathbf{X}), ..., g_s(\mathbf{X})$, a set term V in place of $h(\mathbf{Y}, \mathbf{X})$, and a boolean term U in place of $p(Z, \mathbf{Y}, \mathbf{X})$. We would like also to avoid writing the existential quantifiers, so we need some notation to identify the bound variables. We express all these conditions by the following term formation rule.

BT 6: Assume $\mathrm{V}, \mathrm{W}_1, ..., \mathrm{W}_s$ are set terms, U is a boolean term, the variable Z does not occur free in the terms $\mathrm{V}, \mathrm{W}_1, ..., \mathrm{W}_s$, and the variables $Y_1, ..., Y_s$ do not occur free in the terms $\mathrm{W}_1, ..., \mathrm{W}_s$. The following is a set term:

$$\{Z \in \mathrm{V} : Y_1 \in \mathrm{W}_1 \wedge ... \wedge Y_s \in \mathrm{W}_s : \mathrm{U}\}.$$

The semantics of the term in rule **BT 6** is a bit delicate, as the variables $Z, Y_1, ..., Y_s$ are bound in the construction. We denote the term by P, and $\mathbf{X}$ is a list that contains all the free variables in P (and possibly variables that do not occur free in P). We must determine the set operation f_P such that $f_\mathrm{P}(\mathbf{X}) = \mathrm{P}$. We set $g_1(\mathbf{X}) = \mathrm{W}_1, ..., g_s(\mathbf{X}) = \mathrm{W}_s$, $h(\mathbf{Y}, \mathbf{X}) = \mathrm{V}$, $p(Z, \mathbf{Y}, \mathbf{X}) \equiv \mathrm{U}$. If f is the set operation introduced by the replacement rule, we set $f_\mathrm{P} = f$. It follows that P satisfies the axiom,

$$Z \in \mathrm{P} \equiv (\exists Y_1 \in \mathrm{W}_1)...(\exists Y_s \in \mathrm{W}_s)(Z \in \mathrm{V} \wedge \mathrm{U}).$$

EXAMPLE 2.1 Consider the unary set operation f introduced by replacement in the form,

$$f(X) = \{Z \in X; Y : Y \in X : Z \in Y\}.$$

Here, W is the term X, V is the term $X; Y$, and U is the term $Z \in Y$. It follows that the operation f satisfies the axiom,

$$Z \in f(X) \equiv (\exists Y \in \mathrm{W})(Z \in \mathrm{V} \wedge \mathrm{U}).$$

This axiom holds for arbitrary sets Z and X. □

EXERCISE 2.1 Let f be the set operation in Example 2.1. Prove that the following condition is satisfied: $Z \in f(X) \equiv Z \in X \wedge (\exists Y \in X) Z \in Y$.

A useful simplification is used very often in applications of the rule. We may write $Z = \mathrm{V}$ instead of $Z \in \mathrm{V}$, in which case the official notation would be $Z \in \{\mathrm{V}\}$. For example, we can introduce a unary set operation f in the form

$$f(X) = \{Z = <Y, Y> : Y \in X : X \in X\},$$

which means that the elements of $f(X)$ are all sets of the form $<Y, Y>$ with $Y \in X$ and $X \in X$. Hence, if $X \notin X$, then $f(X) = \emptyset$.

This notation can be simplified a bit more in those cases where the variable Z does not occur free in the boolean condition. If this is the case we can write simply V instead of $Z = \mathrm{V}$. For example, the above set operation f can be expressed in the form

$$f(X) = \{<Y, Y> : Y \in X : X \in X\}.$$

The preceding construction is very important and it is convenient to introduce a new rule of term formation.

BT 7: Assume $\mathrm{V}, \mathrm{W}_1, \ldots, \mathrm{W}_s$ are set terms, U is a boolean term and the variables $Y_1, \ldots, Y_s$ do not occur free in the terms $\mathrm{W}_1, \ldots, \mathrm{W}_s$. The following is a set term:

$$\{\mathrm{V} : Y_1 \in \mathrm{W}_1 \wedge \ldots \wedge Y_s \in W_s : \mathrm{U}\}.$$

In this construction the only bound variables are $Y_1, ..., Y_s$ that do not occur free in the terms $\mathrm{W}_1, ..., \mathrm{W}_s$.

EXAMPLE 2.2 Consider the binary set operation f' introduced by replacement in the form,

$$f'(X,Y) = \{Z \in X; Y :: Y \in X \wedge Z \in Y\}.$$

Note that the pair :: indicates that the set of existential variables is empty (so $s = 0$ in the rule). It follows that the operation f' satisfies the axiom,

$$Z \in f'(X,Y) \equiv Z \in X; Y \wedge Y \in X \wedge Z \in Y.$$

We conclude that $Z \in f'(X,Y) \equiv Z \in X \wedge Y \in X \wedge Z \in Y$. □

The type of replacement discussed in the preceding example, where there is no existential quantifier and $s = 0$ in the rule, is usually called *separation*. On the other hand, the predicate p in the rule can be omitted, and this amounts to assuming a true predicate, say $p(X) \equiv X = X$. This type of replacement we call *pure replacement*. We may even have separation and pure replacement in the same application of the rule. For example, we can set $f(X) = \{Z \in \mathrm{V} ::\}$, which clearly is equivalent to setting $f(X) = \mathrm{V}$. We can also write $f(X) = \{\mathrm{V} ::\}$, and this is equivalent to $f(X) = \{\mathrm{V}\}$.

Remark 2.1 The preceding examples show that the organization determined by the colons in a term under rules **BT 6** or **BT 7** is essential. The two examples are quite similar, but in Example 2.1 the only free variable is X, and Y is bound, while in Example 2.2 both variables are free. □

EXERCISE 2.2 Let g_1 be a unary set operation, g_2 a binary set operation, h a 3-ary set operation, and p a 3-ary set predicate. Introduce, using replacement, a unary set operation f that satisfies the condition:

$$Z \in f(X) \equiv (\exists Y_1 \in g_1(X))(\exists Y_2 \in g_2(Y_1, X))(Z = h(Y_1, Y_2, X) \wedge p(Y_1, Y_2, X)).$$

Remark 2.2 Note that braces are used in this work in two different ways. Definitions 1.6.1 and 1.6.2 require the braces to denote the singleton operation and the general insertion operation which may be considered a generalization of the singleton operation. On the other hand, braces in rules **BT 6** and **BT 7** are used to denote the application of the replacement rule. The latter requires a pair of colons, to make clear which is the intended operation. To avoid confusions we intend to write explicitly all the braces and colons required by rules **BT 6** and **BT 7**, even if in some applications this may appear redundant or unnecessary. Replacement is the crucial mechanism to generate new operations from given operations and predicates, and it is important to make as explicit as possible the applications of the rule. □

EXAMPLE 2.3 Consider the following unary set operation f introduced by separation as follows:

$$f(X) = \{Z \in X :: Z \notin Z\}.$$

If $f(X) \in X$, it follows that $f(X) \in f(X)$ if and only if $f(X) \notin f(X)$, which is a contradiction. We conclude that $f(X) \notin X$, so there is no set X that contains all the sets in the universe. □

We proceed now to use replacement to define a number of important set operations. We assume $k \geq 1$.

2.1.1 $\bigcup X = \{Z \in Y : Y \in X :\}$ (*the general union operation*).

2.1.2 $\bigcap X = \{Z \in Y : Y \in X : (\forall V \in X) Z \in V\}$ (*the general intersection operation*).

2.1.3 $X_1 \setminus X_2 = \{Z \in X_1 :: Z \notin X_2\}$ (*the subtraction operation*).

2.1.4 $X_1 \cup ... \cup X_k = \bigcup\{X_1, ..., X_k\}$ (*finite union operation*).

2.1.5 $X_1 \cap ... \cap X_k = \bigcap\{X_1, ..., X_k\}$ (*finite intersection operation*).

2.1.6 $X_1 \ominus X_2 = X_1 \setminus \{X_2\}$ (*the reduction operation*).

EXERCISE 2.3 Let V,W be set terms and U a boolean term such that the variable Z does not occur free in the terms V, W and the variable Y does not occur free in W (see rules **BT 6** and **BT 7**). Prove that

$$\{Z \in \mathrm{V} : Y \in \mathrm{W} : \mathrm{U}\} = \bigcup\{\mathrm{V} : Y \in \mathrm{W} : \mathrm{U}\}.$$

EXERCISE 2.4 Let X and Z be sets. Prove the following conditions are equivalent:

(a) $Z \in \bigcup X$.

(b) $(\exists Y \in X) Z \in Y$.

EXERCISE 2.5 Let X and Z be sets such that $\bigcup X = Z$. Prove:

(a) $(\forall Y \in X) Y \subseteq Z$.

(b) If Z' is a set such that $(\forall Y \in X) Y \subseteq Z'$, then $Z \subseteq Z'$.

EXERCISE 2.6 Let X be a set and $Z \in X$ be maximal in X, in the sense that whenever $Y \in X$, then $Y \subseteq Z$. Prove that $\bigcup X = Z$.

EXERCISE 2.7 Let Z be a set. Prove the following conditions are equivalent:

(a) Z is transitive.

(b) $\bigcup Z \subseteq Z$.

EXERCISE 2.8 Prove that if a set Z is transitive, then $\bigcup Z$ is also transitive.

EXERCISE 2.9 Assume every element of the set Z is transitive. Prove that $\bigcup Z$ is transitive.

EXERCISE 2.10 Let X and Y be sets. Prove:

(a) If $X = \{Y\}$, then $\bigcup X = Y$.

(b) If $X = S(Y)$ and Y is transitive, then $\bigcup X = Y$.

EXERCISE 2.11 Let X, Y and Z be arbitrary sets. Prove:

(a) $((X \ominus Y) \ominus Z) = ((X \ominus Z) \ominus Y)$.

(b) If $Y \neq Z$, then $((X \ominus Y); Z) = ((X; Z) \ominus Y)$

Remark 2.3 The reader familiar with standard set theory may be interested in the proof of the replacement rule from the standard replacement axiom. We give here a short proof that shows how to derive the rule using also union and separation (usually independent operational axioms in standard set theory). For example, consider a set operation f introduced by the replacement rule:

$$f(X) = \{Z \in h(Y,X) : Y \in g(X) : p(Z,Y,X)\},$$

and we want to prove the axiom,

$$Z \in f(X) \equiv (\exists Y \in g(X))(Z \in h(Y,X) \wedge p(Z,Y,X)).$$

We introduce an auxiliary predicate q such that

$$q(Y,X,V) \equiv V \subseteq h(Y,X) \wedge (\forall Z \in h(Y,X))(Z \in V \equiv p(Z,Y,X)),$$

and by separation it follows that $(\forall Y \in g(X))(\exists! V) q(Y,X,V)$, where the symbol ! indicates that the V is unique. By the standard replacement axiom there is a set W such that $(\forall V)(V \in W \equiv (\exists Y \in g(X)) q(Y,X,V))$. Now we can define the set operation f using separation and union in the form,

$$f(X) = \{Z \in \bigcup W : (\exists Y \in g(X)) p(Z,Y,X)\}.$$

The operation f introduced in this way satisfies the above axiom for the replacement rule.

We can see that the replacement axiom in standard set theory provides an existential bound W for the application of the operational separation axiom. On the other hand, the replacement rule in the system $\mathbb{G}$ is immediately operational without any intermediate bound. □

EXERCISE 2.12 Let X, Y, Z be sets such that $Z \in h(Y, X)$ and $Y \in g(X)$, where h and g are the operations in Remark 2.3 and $p(Z, Y, X)$ holds. Prove $(\exists V \in W)(Z \in V \wedge q(Y, X, V))$.

2.2 Extensional Predicates

The replacement rule introduces a set operation f that is characterized by an axiom of the form,

$$Z \in f(\mathbf{X}) \equiv q(Z, \mathbf{X}),$$

where q is derived from given set operations and predicates using conjunction, disjunction and existential local quantification. Here we generalize the procedure to a more general class of predicates. For this purpose we must identify a special argument among the possibly many arguments of a predicate. We agree that in every case the first argument is chosen with the understanding that the construction will not change if we choose a different one.

Assume q is a $(k+1)$-ary predicate and f is a k-ary set operation. We say that f is an *extensional operation* for q if the condition,

$$Z \in f(\mathbf{X}) \equiv q(Z, \mathbf{X}),$$

is satisfied by all sets $Z, \mathbf{X}$. Note that if f and f' are both extensional operations for the same predicate q we have $f(X) = f'(X)$ for every set X. So the extensional operation for a predicate q is completely determined by q. If there is a general set operation that is extensional for q, we say that q is *extensional*. In Section 3.3 we study the inductive properties of extensional predicates.

EXAMPLE 2.4 Consider the binary predicate $\mathbf{EP}(Z, X) \equiv Z \in X$. We set $f(X) = X$, so f is a projection operation and f is an extensional operation for $\mathbf{EP}$ since we have $Z \in f(X) \equiv \mathbf{EP}(Z, X)$; hence the predicate $\mathbf{EP}$ is extensional. □

EXAMPLE 2.5 We consider now the binary predicate $q(Z, X) \equiv X \in Z$. If f is an extensional operation for q, it follows that $Z \in f(X) \equiv X \in Z$. Note that now we can introduce a unary set operation f' such that

$$f'(X) = f(X) \cup \{Y \ominus X : Y \in f(X) :\},$$

and it follows that $Z \in f'(X)$ for arbitrary sets X and Z. We know this condition is impossible in our system (see Example 2.3). We conclude that if the system $\mathbb{G}$ is consistent, then we cannot define a set operation that is extensional for the predicate q. □

EXERCISE 2.13 Let f' be the set operation in Example 2.5. Prove that if Z is an arbitrary set, then $Z \in f'(\emptyset)$.

EXERCISE 2.14 Assume the predicate q is extensional and the predicate q' satisfies the condition $q'(Z, \mathbf{X}) \to q(Z, \mathbf{X})$. Prove that q' is extensional.

EXAMPLE 2.6 The operation $f(X) = \{X\}$ is an extensional operation for the equality predicate **EQ**, so the equality predicate is extensional. □

Remark 2.4 We are interested in rules that can be used to generate basic predicates, in such a way that whenever q is generated we should be able to determine how the extensional operation for q is generated in the system $\mathbb{G}$. In this way we can be sure of the existence of a set operation satisfying an extensional condition. We proceed to describe a number of rules that satisfy this condition. □

The most important rule is, in fact, the replacement rule that introduces a set operation f with a predicate described in the replacement axiom, where the axiom asserts that f is an extensional operation for the predicate.

Assume f_1 is an extensional operation for the predicate q_1 and f_2 is an extensional operation for the predicate q_2. It follows that $f(\mathbf{X}) = f_1(\mathbf{X}) \cup f_2(\mathbf{X})$ is an extensional operation for the predicate $q(Z, \mathbf{X}) \equiv q_1(z, \mathbf{X}) \vee q_2(Z, \mathbf{X})$.

Concerning conjunction we have a stronger rule in the sense that we do not need both conjuncts to be extensional. Assume h is an extensional operation for the basic predicate q and p is an arbitrary general predicate of the same arity. Using separation we introduce a set operation $f(\mathbf{X}) = \{Z \in h(\mathbf{X}) :: p(Z,\mathbf{X})\}$ and it follows that f is an extensional operation for $q(Z,\mathbf{X}) \wedge p(Z,\mathbf{X})$.

We now consider quantifiers, so let f be an extensional operation for the $(k+2)$-ary predicate q, and let q_1 and q_2 be the $(k+1)$-ary predicates,

$$q_1(Z,\mathbf{X}) \equiv (\exists Y \in g(\mathbf{X}))q(Z,\mathbf{X},Y)$$
$$q_2(Z,\mathbf{X}) \equiv g(\mathbf{X}) \neq \emptyset \wedge (\forall Y \in g(\mathbf{X}))q(Z,\mathbf{X},Y)$$

The following relations follow from these definitions:

$$q_1(Z,\mathbf{X}) \equiv Z \in \bigcup\{f(\mathbf{X},Y) : Y \in g(\mathbf{X}) :\}$$
$$q_2(Z,\mathbf{X}) \equiv Z \in \bigcap\{f(\mathbf{X},Y) : Y \in g(\mathbf{X}) :\}$$

We conclude that q_1 and q_2 are extensional predicates.

We must include a rule of substitution with given set operations, provided the occurrences of the variable Z are not changed by the substitution. For example, if f is an extensional operation for the binary predicate q, then $f'(\mathbf{X}) = f(g(\mathbf{X}))$ is an extensional operation for the $(k+1)$-ary predicate $q'(Z,\mathbf{X}) \equiv q(Z,g(\mathbf{X}))$.

EXAMPLE 2.7 Consider the predicate $q(Z,X) \equiv (\exists V \in X)Z = X \ominus V$. It follows that $q(Z,X) \equiv Z \in \{X \ominus V : V \in X :\}$, so q is extensional. □

2.3 Relations and Functions

We have already defined the ordered pair operation that is the basis of the theory of relations. We need now the decoding operations, to derive the components from the pairing. First we prove a result that relates union, intersection, and subtraction.

Theorem 2.1 *If X and Y are arbitrary sets, then* $Y = ((X \cup Y) \setminus X) \cup (X \cap Y))$.

PROOF. We prove the equation using extensionality. Assume Z is an element of Y (the left side of the equation). It follows immediately that $Z \in X$ or $Z \notin X$, and in both cases Z is in the right side of the equation. Assume now that Z is in the right side of the equation. This induces two cases: $Z \in X \cap Y$ or $Z \in ((X \cup Y) \setminus X)$. In both cases we have $Z \in Y$. □

2.3.1 $[Z]_1 = \bigcup \bigcap Z$ (*the left component operation*).

2.3.2 $[Z]_2 = (\bigcup \bigcup Z \setminus [Z]_1) \cup \bigcap \bigcup Z$ (*the right component operation*).

Theorem 2.2 *If* $Z = <X,Y>$, *then* $[Z]_1 = X$ *and* $[Z]_2 = Y$.

PROOF. From the definition of $<X,Y>$ it follows that $\bigcap Z = \{X\}$, hence $[Z]_1 = \bigcup\{X\} = X$. Furthermore, $\bigcup \bigcup Z = X \cup Y$, and $\bigcap \bigcup Z = X \cap Y$. It follows that $[Z]_2 = ((X \cup Y) \setminus X) \cup (X \cap Y)) = Y$ by Theorem 2.1. □

2.3.3 $\mathbf{OP}(Z) \equiv Z = <[Z]_1, [Z]_2>$ (*Z is an ordered pair*).

2.3.4 $\mathbf{RL}(Z) \equiv (\forall Y \in Z)\mathbf{OP}(Y)$ (*Z is a relation*).

2.3.5 $\mathbf{FU}(Z) \equiv \mathbf{RL}(Z) \wedge (\forall X \in Z)(\forall Y \in Z)([X]_1 \neq [Y]_1 \vee [X]_2 = [Y]_2)$ (*Z is a function*).

2.3.6 $X_1 \times X_2 = \{<Y_1, Y_2> : Y_1 \in X_1 \wedge Y_2 \in X_2 :\}$ (*the cartesian product operation*).

EXAMPLE 2.8 Consider the following application of replacement, where a binary set operation f is introduced by the term,

$$f(X_1, X_2) = \{[Y]_2 : Y \in X_1 : \mathbf{OP}(Y) \wedge [Y]_1 \in X_2\}.$$

The definition means that a set Z is an element of $f(X_1, X_2)$ if and only if there is a set $X \in X_2$ such that $<X, Z> \in X_1$. □

EXAMPLE 2.9 Consider the following "definition by replacement" of a unary set operation f,

$$f(X) = \{<Y_1, Y_2> : Y_1 \in X \wedge Y_2 \in Y_1 :\}.$$

This is not a correct application of replacement, because the range of the variable Y_2 is not given in terms of the input X. An equivalent correct definition is

$$f(X) = \bigcup\{\{<Y_1, Y_2> : Y_2 \in Y_1 :\} : Y_1 \in X :\},$$

where we use the union operation and two nested applications of replacement. □

We introduce a few more set operations that are relevant for the theory of relations and functions.

2.3.7 $\mathbf{do}(R) = \{[Y]_1 : Y \in R :\}$ (*the domain of R*).

2.3.8 $\mathbf{ra}(R) = \{[Y]_2 : Y \in R :\}$ (*the range of R*).

2.3.9 $\mathbf{ap}(F, X) = \bigcup\{[Y]_2 : Y \in F : [Y]_1 = X\}$ (*the application of F to X*).

2.3.10 $F[X] = \{\mathbf{ap}(F, Y) : Y \in X :\}$ (*the image of X under F*).

2.3.11 $\mathbf{FU1}(F) \equiv \mathbf{FU}(F) \wedge (\forall X \in \mathbf{do}(F))(\forall Y \in \mathbf{do}(F))(\mathbf{ap}(F, X) = \mathbf{ap}(F, Y) \rightarrow X = Y)$(*F is a 1–1 function*).

2.3.12 $\mathbf{NET}(F) \equiv \mathbf{FU}(F) \wedge \mathbf{DR}((F)) \wedge (\forall Y \in \mathbf{do}(F))(\forall Z \in \mathbf{do}(F))(Y \subseteq Z \rightarrow \mathbf{ap}(F, Y) \subseteq \mathbf{ap}(F, Z))$ (*F is a net*).

EXERCISE 2.15 Assume F is a net, $Z \in \mathbf{do}(F)$ and $\bigcup \mathbf{do}(F) = Z$. Prove:

(a) If $Y \in \mathbf{do}(F)$, then $Y \subseteq Z$ and $\mathbf{ap}(F, Y) \subseteq \mathbf{ap}(F, Z)$.

(b) $\mathbf{ap}(F, Z) = \bigcup\{\mathbf{ap}(F, Y) : Y \in \mathbf{do}(F) :\}$.

EXERCISE 2.16 Assume F is a function, $X \in \mathbf{do}(F)$, $\mathbf{ap}(F, X) = Y$, and Y' is an arbitrary set. Prove:

(a) $Y \in \mathbf{ra}(F)$.

(b) $<X, Y'> \in F \equiv Y = Y'$.

EXERCISE 2.17 Let F be a 1–1 function. Define a 1–1 function F' such that whenever $Y \in \mathbf{ra}(F)$, then $Y \in \mathbf{do}(F')$, $\mathbf{ap}(F', Y) \in \mathbf{do}(F)$ and $\mathbf{ap}(F, \mathbf{ap}(F', Y)) = Y$.

2.4 Definition by Cases and Local Abstraction

In this section we introduce two new rules derived from the replacement rule. They play important roles in the basic rules of induction and recursion, and it is useful to have an explicit notation to indicate their application. This notation is obtained by new rules of term construction.

Let p be a given k-ary predicate and f_1, f_2 given k-ary set operations. We introduce by replacement a k-ary set operation h as follows:

$$h(\mathbf{X}) = \bigcup(\{f_1(\mathbf{X}) :: p(\mathbf{X})\} \cup \{f_2(\mathbf{X}) :: \neg p(\mathbf{X})\}).$$

Clearly, the operation h satisfies the following conditions:

$h(\mathbf{X}) = f_1(\mathbf{X})$ if $p(\mathbf{X})$,

$h(\mathbf{X}) = f_2(\mathbf{X})$ if $\neg p(\mathbf{X})$.

To describe the set operation h introduced by the rule we shall use the notation,

$$h(\mathbf{X}) = [p(\mathbf{X}) \Rightarrow f_1(\mathbf{X}), f_2(\mathbf{X})].$$

We say that h is introduced by cases from the predicate p and the operations f_1 and f_2.

EXAMPLE 2.10 As given, the rule allows for definitions with only one case (the predicate p). We can easily extend the application to two or more cases as follows. If p_1, p_2 are given set predicates, and f_1, f_2, and f_3 are given set operations, we use the rule twice to introduce a set operation h such that

$$h(\mathbf{X}) = [p_1(\mathbf{X}) \Rrightarrow f_1(\mathbf{X}), [p_2(\mathbf{X}) \Rrightarrow f_2(\mathbf{X}), f_3(\mathbf{X})]],$$

and it is clear that $h(\mathbf{X}) = f_i(\mathbf{X})$ where i is the first index such that $p_i(\mathbf{X})$ holds, or otherwise $i = 3$. □

We introduce a new rule of term construction that corresponds to definitions by cases. Rather than allowing a basic term for just one case, the rule below holds for any number of cases, following the reduction explained in the preceding example.

BT 8: If $\mathrm{U}_1, ..., \mathrm{U}_m$, $m \geq 1$ are boolean terms, and $\mathrm{V}_1, ..., \mathrm{V}_m, \mathrm{V}_{m+1}$ are set terms, then $[\mathrm{U}_1, ..., \mathrm{U}_m \Rrightarrow \mathrm{V}_1, ..., \mathrm{V}_m, \mathrm{V}_{m+1}]$ is also a set term.

The terms $\mathrm{V}_1, ..., \mathrm{V}_m$ in rule **BT 8** are called the *local exit terms*, and the term V_{m+1} is called the *global exit term*.

The semantics for the term U introduced by rule **BT 8** is clear when there is only one case. If there are $m + 1$ cases the semantics is given by the recursive expression,

$$f_{\mathrm{U}}(\mathbf{X}) = [\mathrm{U}_1 \Rrightarrow \mathrm{V}_1, [\mathrm{U}_2, ..., \mathrm{U}_{m+1} \Rrightarrow \mathrm{V}_2, ..., \mathrm{V}_{m+1}, \mathrm{V}_{m+2}]].$$

EXAMPLE 2.11 Using two cases we can define a binary set operation h such that

$$h(X_1, X_2) = [\mathbf{TR}(X_2), X_1 = \emptyset \Rrightarrow S(X_2), S(X_1), \emptyset].$$

Hence, h is a basic binary set operation such that $h(X_1, X_2) = \emptyset$ if X_2 is not transitive and X_1 is non-empty. □

Remark 2.5 When the global exit term is the constant value $\emptyset$ we may omit writing this value explicitly. In this way the definition of the operation h in Example 2.11 can be written as

$$h(X_1, X_2) = [\mathbf{TR}(X_2), X_1 = \emptyset \Rrightarrow S(X_2), S(X_1)].$$

With this notation we use the same number of boolean terms and set terms, and no confusion may arise concerning the format being used. □

We introduce another derived rule that corresponds to the usual mathematical notion of abstraction. By abstraction in set theory we mean a process where from a given set operation h we derive the set of pairs of the form $<Y, f(Y)>$. In order to do this we must impose the condition that $Y \in X$ for some fixed set X. We call this form of abstraction local abstraction.

Formally, given a $(k+1)$-ary set operation h and a k-ary set operation g, we introduce by replacement the k-ary set operation f such that

$$f(\mathbf{X}) = \{<Y, h(Y, \mathbf{X})> : Y \in g(\mathbf{X}) :\},$$

and we say that f is the abstraction of h relative to g.

Our purpose here is to introduce a special notation via terms representing local abstraction. We need this notation because local abstraction plays a role in the rule of set recursion.

BT 9: Let W and V be set terms where the variable Y does not occur free in W. Then $(\lambda Y \in \mathrm{W})\mathrm{V}$ is a set term.

To explain the semantics of this term we shall say only that it is the same as the following set term under rule **BT 7** (see Remark 2.1),

$$\{<Y, \mathrm{V}> : Y \in \mathrm{W} :\}.$$

EXERCISE 2.18 Let V and W be basic set terms and assume the variable Y does not occur free in W. Prove that $F = (\lambda Y \in \mathrm{W})\mathrm{V}$ is a function, $\mathbf{do}(F) = \mathrm{W}$ and $\mathbf{ap}(F, Y) = \mathrm{V}$ whenever $Y \in \mathbf{do}(F)$.

An application of local abstraction is the binary set operation **rt**, which we write $\mathbf{rt}(F, X) = F{\restriction}X$.

2.4.1 $F{\restriction}X = (\lambda Y \in X \cap \mathbf{do}(F))\mathbf{ap}(F, Y)$ (*The restriction of F to X*).

Note that $F{\restriction}X$ is always a function, even if F is not a a function. Furthermore, $\mathbf{do}(F{\restriction}X) = X \cap \mathbf{do}(F)$, and if $Y \in X \cap \mathbf{do}(F)$ then $\mathbf{ap}(F{\restriction}X, Y) = \mathbf{ap}(F, Y)$.

Theorem 2.3 *Assume F is a function. Then*

$$(\lambda Y \in \mathbf{do}(F))\mathbf{ap}(F, Y) = F.$$

PROOF. Consider $<Y, \mathbf{ap}(F, Y)>$ in the left side, where $Y \in \mathbf{do}(F)$. It follows that $<Y, \mathbf{ap}(F, Y)> \in F$. Conversely, assume $<Y, Y'> \in F$, hence $Y \in \mathbf{do}(F)$ and $\mathbf{ap}(F, Y) = Y'$ and $<Y, Y'> \in (\lambda Y \in \mathbf{do}(F))\mathbf{ap}(F, Y)$. □

2.5 Local Universes

The global universe (or universal domain) is by definition closed under all the operations and predicates definable in the system $\mathbb{G}$. On the other hand, the universe is not a well-defined totality, and as a consequence is not a set. In this section we study well-defined totalities (sets) that are closed under several rules, including the fundamental replacement rule. Later we shall give concrete examples of such sets.

We call the rules defined before (empty set, equality, membership, insertion, projection, substitution, boolean connectives, quantifiers, replacement) *elementary rules*, and the operations and predicates that are introduced by the iterative application of elementary rules we call *elementary operations* and *elementary predicates*. Hence, the elementary set operations are basic set operations, and the elementary set predicates are basic set predicates. We conclude that set operations and predicates obtained from elementary set operations and predicates by elementary rules are also elementary. Note that the rules to be introduced later in this work (set induction, set recursion, omega, ordinal permutation) are not elementary and cannot be used to generate elementary operations or predicates.

In particular, a set X is elementary if the 0-ary set operation $f() = X$ is elementary. For example, $\{\{\{\emptyset\}\}\}$ is elementary. In general, if X is elementary, then $\{X\}$ and $S(X)$ are elementary.

Let U be a fixed set. A set operation f (a set predicate p) is *elementary in U* if it is obtained from an elementary set operation (elementary set predicate) by substitution with elements from U. Such an operation (or predicate) is a general operation (predicate) in the sense explained in Chapter 1. Note that predicates and operations obtained by the application of elementary rules to predicates and operations elementary in U are also elementary in U.

EXERCISE 2.19 Let U and X be sets. Prove:

(a) If X is elementary, then X is elementary in U.

(b) If $X \in U$, then X is elementary in U.

(c) If X is elementary in U, then $S(X)$ is elementary in U.

EXERCISE 2.20 Assume q and q' are unary predicates elementary in the set U. Let $p(X) \equiv q(X) \vee q'(X)$. Prove that p is elementary in U.

We introduce several definitions where the notation $\mathbf{X} \in U$ means $X_1 \in U \wedge ... \wedge X_k \in U$, assuming $\mathbf{X}$ is the list $X_1, ..., X_k$.

Let f be a k-ary set operation. We say that *U is closed under f* if whenever $\mathbf{X} \in U$, then $f(\mathbf{X}) \in U$. In this definition f is usually a general set operation defined by the rules of the system $\mathbb{G}$ and, in particular, can be a set operation elementary in U.

A 2-ary predicate p is *local in U* if whenever $Z \in U$ and the condition, $(\forall Y \in Z)(\exists V \in U)p(Y, V)$, is satisfied, then there is a set $W \in U$ and the condition, $(\forall Y \in Z)(\exists V \in W)p(Y, V)$, is satisfied.

Remark 2.6 Assume that in the definition of locality the set $W \in U$ satisfies the condition $(\forall Y \in Z)(\exists V \in W)p(Y,V)$. Then, any set $W' \in U$ such that $W \subseteq W'$ satisfies the same condition. □

If q is a $(k+1)$-ary predicate we can define a new k-ary predicate p in the form,

$$p(\mathbf{X}) \equiv (\exists V \in U)q(V, \mathbf{X}).$$

In this definition we are first using existential local quantification and then substitution with U. We say that p is obtained from q by *existential quantification over U*. If q is elementary in U and $U \in U$, then p is elementary in U.

A k-ary predicate p is *Σ-elementary over U* if there is a $(k+1)$-ary predicate q such that q is elementary in U and the following relation is satisfied for all $\mathbf{X} \in U$:

$$p(\mathbf{X}) \equiv (\exists V \in U)q(V, \mathbf{X}).$$

Note that the condition may not be satisfied if the components of $\mathbf{X}$ are not elements of U.

EXERCISE 2.21 Let p be a $(k+1)$-ary predicate. Prove:

(a) If p is elementary in U, then p is Σ-elementary over U.

(b) If p is obtained from q by existential quantification over U and q is elementary in U, then p is Σ-elementary over U.

We say that the set $\mathtt{U}$ is a *local universe* if the following conditions are satisfied:

LU 1 $\mathtt{U}$ is transitive and well-founded.

LU 2 $\mathtt{U}$ is closed under every elementary set operation.

LU 3 Every 2-ary predicate elementary in U is local in U.

From now on U denotes a fixed local universe.

EXAMPLE 2.12 Assume $X \in \mathtt{U}$. It follows that $\{X\} \in \mathtt{U}$, $\{\{X\}\} \in \mathtt{U}$, etc. Informally, we conclude that U cannot be finite. □

EXAMPLE 2.13 Assume $X \in \mathtt{U}$. Then $[X]_1 \in \mathtt{U}$ and $[X]_2 \in \mathtt{U}$. Hence, if $<Y, Z> \in \mathtt{U}$, then both $Y \in \mathtt{U}$ and $Z \in \mathtt{U}$. □

From the definition we know that if U is a local universe, then U is closed under the elementary operations. This property extends immediately to set operations that are elementary in U. On the other hand, the locality property **LU 3** is assumed explicitly for predicates that are elementary in U. Note that the property applies to $(k+2)$-ary predicates elementary in U, provided that the extra k arguments are assumed to be elements of U.

Theorem 2.4 *Assume p is a binary predicate elementary in* U *and $Z \in \mathtt{U}$ satisfies the condition, $(\forall Y \in Z)(\exists V \in \mathtt{U})p(Y, V)$. There is a relation $R \in \mathtt{U}$ such that $\mathbf{do}(R) = Z$ and whenever $<Y, V> \in R$, then $p(Y, V)$ holds.*

PROOF. Since p is local there is $W \in \mathtt{U}$ such that $(\forall Y \in Z)(\exists V \in W)p(Y, V)$. Hence we can define by replacement $R = \{<Y, V> : Y \in Z \wedge V \in W : p(Y, V)\}$. □

EXERCISE 2.22 Let p be a binary predicate elementary in U that satisfies the condition, $(\forall Y \in Z)(\exists! V \in \mathtt{U})p(Y, V)$, where the symbol ! means that the $V \in \mathtt{U}$ is unique. Prove that there is a function $F \in \mathtt{U}$ such that $\mathbf{do}(F) = Z$ and for every $Y \in Z$ satisfies the condition, $p(Y, \mathbf{ap}(F, Y))$.

Theorem 2.5 *Let p be a 3-ary predicate elementary in* U. *If $Z \in \mathtt{U}$, then*

$$(\forall Y \in Z)(\exists V \in \mathtt{U})p(Y, Z, V) \equiv (\exists W \in \mathtt{U})(\forall Y \in Z)(\exists V \in W)p(Y, Z, V).$$

PROOF. The implication from right to left follows by the transitivity of $\mathbb{U}$. In the other direction assume that the left side holds, and with $Z \in \mathbb{U}$ fixed introduce the binary predicate p' such that

$$p'(Y, V) \equiv p(Y, Z, V).$$

Since p' is elementary in $\mathbb{U}$ and local, the right side follows. □

Theorem 2.6 *Let p' be a k-ary predicate obtained by existential quantification over $\mathbb{U}$ from a predicate p that is Σ-elementary over $\mathbb{U}$. Then p' is Σ elementary over $\mathbb{U}$.*

PROOF. From the assumptions it follows that there is a $(k+2)$-ary predicate q elementary in $\mathbb{U}$ such that

$$p'(\mathbf{X}) \equiv (\exists V \in \mathbb{U})(\exists V' \in \mathbb{U})q(V', V, \mathbf{X})$$

holds whenever $\mathbf{X} \in \mathbb{U}$. We conclude that the relation

$$p'(\mathbf{X}) \equiv (\exists V \in \mathbb{U})q([V]_1, [V]_2, \mathbf{X})$$

also holds whenever $\mathbf{X} \in \mathbb{U}$. It follows that p' is Σ-elementary over $\mathbb{U}$. □

Theorem 2.7 *Let p be a predicate Σ-elementary over $\mathbb{U}$ and $Z \in \mathbb{U}$ a set such that $(\forall Y \in Z)(\exists V \in \mathbb{U})p(Y, V)$ holds. There is a set $W \in \mathbb{U}$ such that the following conditions are satisfied:*

(i) $(\forall Y \in Z)(\exists V \in W)p(Y, V)$.

(ii) $(\forall V \in W)(\exists Y \in Z)p(Y, V)$.

PROOF. From the assumptions it follows that there is a 3-ary predicate q elementary in $\mathbb{U}$ such that the following condition is satisfied:

$$(\forall Y \in Z)(\exists V \in \mathbb{U})(\exists V' \in \mathbb{U})q(V', Y, V).$$

We conclude that there is $W' \in \mathtt{U}$ such that

$$(\forall Y \in Z)(\exists V \in W')q([V]_1, Y, [V]_2),$$

hence if we take $W = \{[V]_2 : V \in W' : (\exists Y \in Z)p(Y, [V]_2)\}$ it follows that $W \in \mathtt{U}$ satisfies the requirements of the theorem. □

Corollary 3.7.1 If p is a predicate Σ-elementary over $\mathtt{U}$, then p is local in $\mathtt{U}$.

PROOF. Immediate from Theorem 2.7 and the definition. □

Remark 2.7 A local universe is a fairly complicated structure, and in order to generate concrete examples we must wait until stronger techniques are available (see Sections 7.3 and A.3). Both closure and locality are completeness properties that require special sets to be elements of the local universe $\mathtt{U}$. Closure is a natural condition that occurs frequently in mathematics, and it is relatively easy to satisfy. In particular, the closure property is preserved under substitution of operations and closure can be generated by a process that systematically inserts the sets required by the condition. On the other hand, locality is not necessarily preserved under substitution or the logical rules. In general, to generate locality we need a transfinite process that in some cases has to be extended to a very high limit. But locality is the heart of a local universe, and in this section we show several important applications (see Examples 2.14 and 2.15). □

Theorem 2.8 *The predicates Σ-elementary over* $\mathtt{U}$ *are closed under substitution with set operations elementary in* $\mathtt{U}$*, disjunction, conjunction, existential local quantification, universal local quantification, and existential quantification over* $\mathtt{U}$*.*

PROOF. Closure under substitution follows from the substitution rule for elementary predicates. Closure under disjunction follows by exportation of existential quantifiers. Closure under conjunction also follows by exportation of quantifiers and contraction of quantifiers as in Theorem 2.6. Closure under existential local quantification follows by permutation of existential quantifiers. Closure under universal local quantification follows from Theorem 2.5. Closure under existential quantification over $\mathtt{U}$ follows from Theorem 2.6. □

EXAMPLE 2.14 In this example we consider a binary Σ-elementary over $\mathtt{U}$ predicate p, and introduce by separation the unary set operation f such that

$$f(Z) = \{Y \in Z :: p(Y, Z)\}.$$

Since f is not elementary in $\mathtt{U}$ we cannot conclude that $f(Z) \in \mathtt{U}$ whenever $Z \in \mathtt{U}$. Still, we know that p satisfies the condition $p(Y, Z) \equiv (\exists V \in \mathtt{U})q(V, Y, Z)$ for $Y, Z \in \mathtt{U}$, where q is a predicate elementary in $\mathtt{U}$. So we introduce a 2-ary set operation f' such that

$$f'(Z, W) = \{Y \in Z :: (\exists V \in W)q(V, Y, Z)\}$$

and note that f' is elementary in $\mathtt{U}$, so $f'(Z, W) \in \mathtt{U}$ whenever $Z, W \in \mathtt{U}$.

We introduce new assumptions as follows. First, we assume there is another binary Σ-elementary over $\mathtt{U}$ predicate p' that satisfies the condition $p'(Y, Z) \equiv (\exists V \in \mathtt{U})q'(V, Y, Z)$ for $Y, Z \in \mathtt{U}$, where q' is elementary in $\mathtt{U}$. Second, we assume that Z is an element of $\mathtt{U}$ that satisfies the condition,

$$(\forall Y \in Z)(p(Y, Z) \equiv \neg p'(Y, Z)).$$

We shall show that under these assumptions it follows that $f(Z) \in \mathtt{U}$. In fact, it is clear that the following condition is satisfied by the predicates q and q':

$$(\forall Y \in Z)(\exists V \in \mathtt{U})(q(V, Y, Z) \vee q'(V, Y, Z)),$$

hence by locality there is a set $W \in \mathtt{U}$ such that

$$(\forall Y \in Z)(p(Y, Z) \equiv (\exists V \in W)q(V, Y, Z)).$$

We conclude that $f(Z) = f'(Z, W)$, so $f(Z) \in \mathtt{U}$. Note that the set $W \in \mathtt{U}$ depends on the predicate p' and the set Z. □

EXAMPLE 2.15 Assume p is a binary predicate Σ-elementary over $\mathtt{U}$, and q is the 3-ary predicate elementary in $\mathtt{U}$ such that the relation $p(Y,V) \equiv (\exists X \in \mathtt{U})q(X,Y,V)$ holds for all $Y,V \in \mathtt{U}$. We introduce a binary elementary set operation f in the form

$$f(Z',W) = \{<Y,V> : Y \in Z' \wedge V \in W : (\exists X \in W)q(X,Y,V)\}.$$

Clearly, if $Z' \in \mathtt{U}$ and $W \in \mathtt{U}$, then $f(Z',W) \in \mathtt{U}$. Furthermore, if $<Y,V> \in f(Z',W)$, $Z' \in \mathtt{U}$ and $W \in \mathtt{U}$, then $Y \in \mathtt{U}$, $V \in \mathtt{U}$, and $p(Y,V)$ holds.

We shall show that, under some assumptions about the set Z' and the predicate p, there is $W \in \mathtt{U}$ such that whenever $Y \in Z'$, $V \in \mathtt{U}$ and $p(Y,V)$ holds, then $<Y,V> \in f(Z',W)$.

We assume that $Z' \in \mathtt{U}$ and furthermore that $(\forall Y \in Z')(\exists! V \in \mathtt{U})p(Y,V)$, where the symbol ! means that the V is unique in $\mathtt{U}$. Since p is local in $\mathtt{U}$ it follows that there is a set $W_1 \in \mathtt{U}$ such that $(\forall Y \in Z')(\exists! V \in W_1)p(Y,V)$.

Next, we define a new binary predicate p' in the form $p'(Y,V) \equiv (\exists V' \in \mathtt{U})(p(Y,V') \wedge V \neq V')$. From the single-valued property of p it follows that $p(Y,V) \equiv \neg p'(Y,V)$ whenever $Y \in Z'$ and $V \in \mathtt{U}$. Furthermore, the predicate p' is Σ-elementary over $\mathtt{U}$ (Theorem 2.8), hence there is a 3-ary predicate q' elementary in $\mathtt{U}$ that whenever $Y,V \in \mathtt{U}$ satisfies the relation,

$$p'(Y,V) \equiv (\exists X \in \mathtt{U})q'(X,Y,V).$$

From Example 2.14 we conclude that $(\forall Y \in Z')(\exists X \in \mathtt{U})(q(X,Y,V) \vee q'(X,Y,V))$ whenever $V \in \mathtt{U}$. Using locality we can replace the existential quantifier with $\exists X \in W_2$, where $W_2 \in \mathtt{U}$. From this we conclude that

$$(\forall Y \in Z')(p(Y,V) \equiv (\exists X \in W_2)q(X,Y,V))$$

holds for all $V \in \mathtt{U}$. Returning to the definition of the set operation f it follows that whenever $Y \in Z'$, $V \in \mathtt{U}$ and $p(Y,V)$ hold, then $<Y,V> \in f(Z',W_1 \cup W_2)$.

The final conclusion is that there is $W \in \mathtt{U}$ such that $f(Z',W) \in \mathtt{U}$ is a function, and for all $Y \in Z'$ and $V \in \mathtt{U}$ the following equivalence holds:

$$p(Y,V) \equiv \mathbf{ap}(f(Z',W),Y) = V.$$

This result is applied in Chapter 5 where we prove that local universes are closed under elementary recursive operations. □

Remark 2.8 Example 2.14 deals with a set operation f obtained by separation from a predicate p that is not elementary in $\mathtt{U}$ and we show that, under some assumptions on p and the set $Z \in \mathtt{U}$, we have $f(Z) \in \mathtt{U}$. The construction in Example 2.15, that depends on Example 2.14, is different because we are dealing with a binary operation f obtained by replacement from a predicate p that, again, is not elementary in $\mathtt{U}$. Here we show that, under some assumptions on Z' and p, there is a set $W \in \mathtt{U}$ such that $f(Z', W)$ is complete relative to the predicate p and $f(Z, W) \in \mathtt{U}$. So in both examples we deal with a problem of closure in a local universe $\mathtt{U}$. In Chapter 5 these constructions are used to prove closure under recursive operations. □

EXERCISE 2.23 Let p be a predicate Σ-elementary over $\mathtt{U}$, and $Z' \in \mathtt{U}$ a set such that $(\forall Y \in Z')(\exists V \in \mathtt{U})p(Y, V)$. Prove there is a function $F \in \mathtt{U}$ such that:

(a) $(\forall Y \in Z')\mathbf{ap}(F, Y) \neq \emptyset$.

(b) $(\forall Y \in Z')(\forall V \in \mathbf{ap}(F, Y))p(Y, V)$.

2.6 Notes

The well-known axiom of replacement is an essential element of standard set theory, although in this frame the motivation is usually given in terms of the "size" of sets (see Kunen (1980) and Hallett (1984)) On the other hand, the primitive construction that supports our replacement rule has no such connotations and simply refers to the fact that the construction is objectively clear and completely determined by the inputs of the operation, hence is independent of the universe.

In Remark 2.3 we show that the replacement rule proposed in this chapter can be derived in standard set theory via the standard replacement axiom, the separation rule, and the union operation. On the other hand, the advantage of the replacement rule is that it is immediately operational. We do not claim that the replacement rule is equivalent to the standard replacement axiom or that it is the only operational version of the axiom. Still, it is sufficient for all the applications in this work, usually in the format provided by rules **BT 6** and **BT 7**.

Notwithstanding the power of the replacement rule, particularly in combination with the rules of Chapter 1, at this stage of the theory we can only generate finite sets. This does not mean, of course, that the axioms imply that the universe contains only finite sets. In fact, later we define by induction the property of being finite, and furthermore we introduce a rule that generates infinite sets. In general, there is no rule in the system that imposes restrictions on what we may want to consider a set in the universe, provided that the principle of extensionality is satisfied. In particular, the standard foundation axiom is not a part of the theory.

The theory of local universes is derived from the theory of admissible sets in Barwise (1975). We use a different notation because there are substantial differences, and there is no reason to identify the two theories. Still, we follow Barwise (1975) closely, in particular in the constructions in Examples 2.14 and 2.15. The principle of reflection, crucial in Barwise's approach, can be avoided because we have a general rule of predicate substitution. This is one of the many advantages of working in a system where operations are primitive. On the other hand, operations create problems in dealing with models of set theory, and this is an important consideration in the theory of admissible sets.

The replacement rule is a general construction whereby new sets can be generated from special combinations of operations and predicates. The notion of extensional predicate provides a generalization that applies to more general combinations. In Section 3.3 we show that extensional predicates induce a natural notion of well-foundedness.

The rule of local abstraction is more definite than the general rule of replacement, for it allows functions to be generated from operations. This transition from an operation (which is not a set) to a function (which is a set) is crucial in set theory. On the other hand, the process involves a real mutilation, for the domain of a function is a set, while the domain of an operation is the universe.

The whole process of recursion, as defined in Chapter 5, depends on local abstraction. In fact, we may consider recursion as an inversion of abstraction, where a set operation is able to reflect on its own abstraction.

Chapter 3

Set Induction

In this chapter we extend the class of basic rules with a new rule of set induction. To some extent this rule is familiar in the standard presentations of set theory, via the so-called axiom of foundation. Here it is taken as a primitive rule that, as with every rule in this work, is supported by a primitive construction to be discussed later in this chapter.

In the Appendix we shall show that inductive predicates can be introduced via explicit definitions involving the enumeration rule. Still, recursive operations cannot be derived from such definitions, due to the predicative frame we impose on inductive proofs.

3.1 Induction Rule

Formally, the rule of set induction involves two given set predicates and one given set operation, and it introduces a new set predicate and a new set operation that satisfy four axioms. When the rule is applied with basic predicates and operations the result is a basic set predicate and a basic set operation.

The Set Induction Rule Assume q_1 and q_2 are $(k+1)$-ary general set predicates, $k \geq 0$, and g is a $(k+1)$-ary general set operation. We introduce a $(k+1)$-ary set predicate ind (*the inductive predicate*) and a $(k+1)$-ary set operation $\mathbf{tc}_{\mathsf{ind}}$ (*the transitive* ind-*closure operation*) that satisfy the following axioms:

The Induction Axiom:

$$\mathsf{ind}(Z,\mathbf{X}) \equiv (q_1(Z,\mathbf{X}) \vee (q_2(Z,\mathbf{X}) \wedge (\forall Y \in g(Z,\mathbf{X}))\mathsf{ind}(Y,\mathbf{X}))).$$

The Closure Axiom:

$$\mathbf{tc}_{\mathsf{ind}}(Z,\mathbf{X}) = [\mathsf{ind}(Z,\mathbf{X}) \Rrightarrow \mathbf{is}_{\mathsf{ind}}(Z,\mathbf{X}) \cup \bigcup\{\mathbf{tc}_{\mathsf{ind}}(Y,\mathbf{X}) : Y \in \mathbf{is}_{\mathsf{ind}}(Z,\mathbf{X}) :\}],$$

where $\mathbf{is}_{\mathsf{ind}}(Z,\mathbf{X}) = \{Y \in g(Z,\mathbf{X}) :: \neg q_1(Z,\mathbf{X}) \wedge q_2(Z,\mathbf{X})\}$; note that the implicit exit value is $\emptyset$.

The Foundation Axiom:

$$(\exists Z \in V)\mathsf{ind}(Z,\mathbf{X}) \rightarrow (\exists Z' \in V)(\mathsf{ind}(Z',\mathbf{X}) \wedge V \cap \mathbf{tc}_{\mathsf{ind}}(Z',\mathbf{X}) = \emptyset).$$

The Transitive Axiom:

$$(\forall Y \in \mathbf{tc}_{\mathsf{ind}}(Z,\mathbf{X}))\mathbf{is}_{\mathsf{ind}}(Y,\mathbf{X}) \subseteq \mathbf{tc}_{\mathsf{ind}}(Z,\mathbf{X}).$$

We say that the predicate ind and the operation $\mathbf{tc}_{\mathsf{ind}}$ are *defined by* ind-*induction* (or *inductively defined*). The variable Z is the *induction variable.* The variables $\mathbf{X}$ are *parameters* in the induction.

We can simplify the axioms, using abbreviations. For example, we set $\mathsf{ind}^*(Z,\mathbf{X}) \equiv q_1(Z,\mathbf{X}) \vee (q_2(Z,\mathbf{X}) \wedge (\forall Y \in g(Z,\mathbf{X}))\mathsf{ind}(Y,\mathbf{X}))$. Now we can write the induction axiom in the form

$$\mathsf{ind}(Z,\mathbf{X}) \equiv \mathsf{ind}^*(Z,\mathbf{X}).$$

Concerning the closure axiom, we have already used an abbreviation in the form of the operation $\mathbf{is}_{\mathsf{ind}}$. Since this operation occurs frequently in this work, we introduce it by a formal definition.

3.1.1 $\mathbf{is}_{\mathsf{ind}}(Z, \mathbf{X}) = \{Y \in g(Z, \mathbf{X}) :: \neg q_1(Z, \mathbf{X}) \wedge q_2(Z, \mathbf{X})\}$.

Note that if $q_1(Z, \mathbf{X}) \vee \neg q_2(Z, \mathbf{X})$, then $\mathbf{is}_{\mathsf{ind}}(Z, \mathbf{X}) = \emptyset$. Otherwise, we have $\mathbf{is}_{\mathsf{ind}}(Z, \mathbf{X}) = g(Z, \mathbf{X})$. In any case, the operation $\mathbf{is}_{\mathsf{ind}}$ is completely independent of the predicate ind. In many cases we refer to the elements of $\mathbf{is}_{\mathsf{ind}}(Z, \mathbf{X})$ as *immediate successors of* Z at $\mathbf{X}$.

The closure axiom can be abbreviated by omitting the condition $\mathsf{ind}(Z, \mathbf{X})$. So we write the axiom in the form,

$$\mathbf{tc}_{\mathsf{ind}}(Z, \mathbf{X}) = \mathbf{is}_{\mathsf{ind}}(Z, \mathbf{X}) \cup \bigcup \{\mathbf{tc}_{\mathsf{ind}}(Y, \mathbf{X}) : Y \in \mathbf{is}_{\mathsf{ind}}(Z, \mathbf{X}) :\},$$

with the understanding that in case $\mathsf{ind}(Z, \mathbf{X})$ fails, then $\mathbf{tc}_{\mathsf{ind}}(Z, \mathbf{X}) = \emptyset$. In applications, the evaluation of $\mathbf{tc}_{\mathsf{ind}}(Z, \mathbf{X})$ can be simplified in several ways. For example, if $q_1(Z, \mathbf{X})$ holds, then $\mathbf{is}_{\mathsf{ind}}(Z, \mathbf{X}) = \mathbf{tc}_{\mathsf{ind}}(Z, \mathbf{X}) = \emptyset$. On the other hand, if $\mathsf{ind}(Z, \mathbf{X}) \wedge \neg q_1(Z, \mathbf{X})$ holds, then $\mathbf{is}_{\mathsf{ind}}(Z, \mathbf{X}) = g(Z, \mathbf{X})$, and the axiom can be written taking this situation into account.

We shall introduce the predicate ind in applications simply by writing the induction axiom, since this determines the other axioms.

EXAMPLE 3.1 Assume $q_1(Z) \equiv Z = \emptyset$, $q_2(Z) \equiv \emptyset \in Z$ and $g(Z) = Z$. We define inductively the set predicate ind such that:

$$\mathsf{ind}(Z) \equiv (q_1(Z) \vee (q_2(Z) \wedge (\forall Y \in g(Z))\mathsf{ind}(Y))).$$

Using the definitions, we can rewrite this axiom in the form,

$$\mathsf{ind}(Z) \equiv Z = \emptyset \vee (\emptyset \in Z \wedge (\forall Y \in Z)\mathsf{ind}(Y)).$$

From this we conclude $\mathsf{ind}(\emptyset)$, and from this $\mathsf{ind}(\{\emptyset\})$. Using the axiom again, we get $\mathsf{ind}(\{\emptyset, \{\emptyset\}\})$. In general, if $\mathsf{ind}(X)$, then $\mathsf{ind}(\{\emptyset, X\})$. Note that $\mathbf{is}_{\mathsf{ind}}(Z) = \{Y \in Z :: Z \neq \emptyset \wedge \emptyset \in Z\}$. Hence $\mathbf{is}_{\mathsf{ind}}(Z) = Z$ if $\emptyset \in Z$ and $\mathbf{is}_{\mathsf{ind}}(Z) = \emptyset$ if $\emptyset \notin Z$. In general, we have $\mathsf{ind}(Z) \rightarrow \mathbf{is}_{\mathsf{ind}}(Z) = Z$; but we may have $\mathbf{is}_{\mathsf{ind}}(Z) = Z$ even if $\neg\mathsf{ind}(Z)$. □

EXERCISE 3.1 Assume $\mathsf{ind}(Z)$, where ind is the inductive predicate in Example 3.1. Prove:

(a) $(\forall Y \in Z)\mathsf{ind}(Y)$.

(b) $\mathbf{tc}_{\mathsf{ind}}(Z) = Z \cup \bigcup\{\mathbf{tc}_{\mathsf{ind}}(Y) : Y \in Z :\}$.

Remark 3.1 The induction rule can be applied when q_1, q_2, g are general rather than basic predicates. In this case the predicate ind is also general. To show this, assume $\mathbf{V} = V_1, V_2, ..., V_s$ are all the sets required to write q_1, q_2, g as substitutions in basic q_1', q_2', g'. We write a new induction for a basic predicate ind′ such that

$$\mathsf{ind}'(Z,\mathbf{X},\mathbf{Y}) \equiv (q_1'(Z,\mathbf{X},\mathbf{Y}) \vee (q_2'(Z,\mathbf{X},\mathbf{Y}) \wedge (\forall Y \in g'(Z,\mathbf{X},\mathbf{Y}))\mathsf{ind}'(Y,\mathbf{X},\mathbf{Y}))),$$

where $\mathbf{Y} = Y_1, ..., Y_s$ are new variables. We can show now that $\mathsf{ind}(Z,\mathbf{X}) \equiv \mathsf{ind}'(Z,\mathbf{X},\mathbf{V})$, so ind is a general predicate. □

EXERCISE 3.2 Let ind and ind′ be the predicates in Remark 3.1. Prove that $\mathsf{ind}(Z,\mathbf{X}) \equiv \mathsf{ind}'(Z,\mathbf{X},\mathbf{V})$ holds for arbitrary sets $Z, \mathbf{X}$.

Theorem 3.1 *Assume* $\mathsf{ind}(Z,\mathbf{X})$. *Then:*

(i) $q_1(Z,\mathbf{X}) \vee q_2(Z,\mathbf{X})$.

(ii) $q_1(Z,\mathbf{X}) \to \mathbf{is}_{\mathsf{ind}}(Z,\mathbf{X}) = \mathbf{tc}_{\mathsf{ind}}(Z,\mathbf{X}) = \emptyset$.

(iii) $\neg q_1(Z,\mathbf{X}) \to q_2(Z,\mathbf{X}) \wedge \mathbf{is}_{\mathsf{ind}}(Z,\mathbf{X}) = g(Z,\mathbf{X})$.

(iv) $(\forall Y \in \mathbf{is}_{\mathsf{ind}}(Z,\mathbf{X}))\mathsf{ind}(Y,\mathbf{X})$.

(v) $\mathbf{is}_{\mathsf{ind}}(Z,\mathbf{X}) \subseteq \mathbf{tc}_{\mathsf{ind}}(Z,\mathbf{X})$.

(vi) $(\forall Y \in \mathbf{is}_{\mathsf{ind}}(Z,\mathbf{X}))\mathbf{tc}_{\mathsf{ind}}(Y,\mathbf{X}) \subseteq \mathbf{tc}_{\mathsf{ind}}(Z,\mathbf{X})$.

(vii) $Z \notin \mathbf{tc}_{\mathsf{ind}}(Z,\mathbf{X})$.

PROOF. Part (i) is clear from the induction axiom. To prove (ii), note that $\mathbf{is}_{\mathsf{ind}}(Z, \mathbf{X}) = \emptyset$ follows from Definition 3.1.1. Part (iii) follows, because from $\neg q_1(Z, \mathbf{X}) \wedge \mathsf{ind}(Z, \mathbf{X})$ and the induction axiom it follows $q_2(Z, \mathbf{X})$, hence from Definition 3.1.1 we have $\mathbf{is}_{\mathsf{ind}}(Z, \mathbf{X}) = g(Z, \mathbf{X})$. Part (iv) is clear if $q_1(Z, \mathbf{X})$, because $\mathbf{is}_{\mathsf{ind}}(Z, \mathbf{X}) = \emptyset$. Otherwise, we use part (iii) with the induction axiom. Parts (v) and (vi) are clear from the closure axiom. To prove (vii) we use the foundation axiom with $V = \{Z\}$ and it follows that $V \cap \mathbf{tc}_{\mathsf{ind}}(Z, \mathbf{X}) = \emptyset$, so $Z \notin \mathbf{tc}_{\mathsf{ind}}(Z, \mathbf{X})$. □

EXAMPLE 3.2 Assume $Y, Z, \mathbf{X}$ are sets such that $Y \in \mathbf{tc}_{\mathsf{ind}}(Z, \mathbf{X})$. This implies $\mathsf{ind}(Z, \mathbf{X})$, otherwise $\mathbf{tc}_{\mathsf{ind}}(Z, \mathbf{X}) = \emptyset$. Consider the set $V = \{Y, Z\}$. By the foundation axiom there is $Z' \in V$ such that $\mathsf{ind}(Z', \mathbf{X}) \wedge V \cap \mathbf{tc}_{\mathsf{ind}}(Z', \mathbf{X}) = \emptyset$. Since $Y \in V \cap \mathbf{tc}_{\mathsf{ind}}(Z, \mathbf{X})$, it follows that $Z' = Y$, so we conclude $\mathsf{ind}(Y, \mathbf{X})$. Hence, if Z is any set, then $(\forall Y \in \mathbf{tc}_{\mathsf{ind}}(Z, \mathbf{X}))\mathsf{ind}(Y, \mathbf{X})$. □

EXERCISE 3.3 Assume $Y, Z, \mathbf{X}$ are sets such that $Y \in \mathbf{tc}_{\mathsf{ind}}(Z, \mathbf{X})$. Prove that $Z \notin \mathbf{tc}_{\mathsf{ind}}(Y, \mathbf{X})$.

EXERCISE 3.4 Assume $W, Y, Z, \mathbf{X}$ are sets such that $Y \in \mathbf{tc}_{\mathsf{ind}}(Z, \mathbf{X}) \wedge W \in \mathbf{tc}_{\mathsf{ind}}(Y, \mathbf{X})$. Prove that $Z \notin \mathbf{tc}_{\mathsf{ind}}(W, \mathbf{X})$.

Theorem 3.2 *Let $Y_0, \mathbf{X}$ be fixed sets. Then:*

(i) *If Z is a set such that $Y_0 \in \mathbf{tc}_{\mathsf{ind}}(Z, \mathbf{X})$ and $\mathbf{tc}_{\mathsf{ind}}(Y_0, \mathbf{X}) \not\subseteq \mathbf{tc}_{\mathsf{ind}}(Z, \mathbf{X})$, then there is $Y \in \mathbf{is}_{\mathsf{ind}}(Z, \mathbf{X})$ such that $Y_0 \in \mathbf{tc}_{\mathsf{ind}}(Y, \mathbf{X}) \wedge \mathbf{tc}_{\mathsf{ind}}(Y_0, \mathbf{X}) \not\subseteq \mathbf{tc}_{\mathsf{ind}}(Y, \mathbf{X})$.*

(ii) *If Z is a set such that $Y_0 \in \mathbf{tc}_{\mathsf{ind}}(Z, \mathbf{X})$, then $\mathbf{tc}_{\mathsf{ind}}(Y_0, \mathbf{X}) \subseteq \mathbf{tc}_{\mathsf{ind}}(Z, \mathbf{X})$.*

PROOF. In part (i) we note that if $Y_0 \in \mathbf{tc}_{\mathsf{ind}}(Z, \mathbf{X})$, then by the closure axiom there are two cases: $Y_0 \in \mathbf{is}_{\mathsf{ind}}(Z, \mathbf{X})$ or $Y_0 \in \mathbf{tc}_{\mathsf{ind}}(Y, \mathbf{X})$ where $Y \in \mathbf{is}_{\mathsf{ind}}(Z, \mathbf{X})$. If we assume $\mathbf{tc}_{\mathsf{ind}}(Y_0, \mathbf{X}) \not\subseteq \mathbf{tc}_{\mathsf{ind}}(Z, \mathbf{X})$, the first case is impossible, so the second case holds and $\mathbf{tc}_{\mathsf{ind}}(Y_0, \mathbf{X}) \not\subseteq \mathbf{tc}_{\mathsf{ind}}(Y, \mathbf{X})$ because $\mathbf{tc}_{\mathsf{ind}}(Y, \mathbf{X}) \subseteq \mathbf{tc}_{\mathsf{ind}}(Z, \mathbf{X})$. Note that $\mathsf{ind}(Y, \mathbf{X})$ follows from Example

3.2. To prove part (ii) we assume $Y_0 \in \mathbf{tc}_{\mathsf{ind}}(Z, \mathbf{X})$ and to get a contradiction we assume that $\mathbf{tc}_{\mathsf{ind}}(Y_0, \mathbf{X}) \not\subseteq \mathbf{tc}_{\mathsf{ind}}(Z, \mathbf{X})$. We introduce a set

$$V = \{Y \in \mathbf{tc}_{\mathsf{ind}}(Z, \mathbf{X}) :: Y_0 \in \mathbf{tc}_{\mathsf{ind}}(Y, \mathbf{X}) \wedge \mathbf{tc}_{\mathsf{ind}}(Y_0, \mathbf{X}) \not\subseteq \mathbf{tc}_{\mathsf{ind}}(Y, \mathbf{X})\}.$$

From part (i) and the foundation axiom there is $Z' \in V$ such that $V \cap \mathbf{tc}_{\mathsf{ind}}(Z', \mathbf{X}) = \emptyset$. By part (i) applied to Z' there is $Y' \in \mathbf{is}_{\mathsf{ind}}(Z', \mathbf{X})$ such that $Y_0 \in \mathbf{tc}_{\mathsf{ind}}(Y', \mathbf{X}) \wedge \mathbf{tc}_{\mathsf{ind}}(Y_0, \mathbf{X}) \not\subseteq \mathbf{tc}_{\mathsf{ind}}(Y', \mathbf{X})$. By the transitive axiom we have $Y' \in \mathbf{tc}_{\mathsf{ind}}(Z, \mathbf{X})$, so $Y' \in V \cap \mathbf{tc}_{\mathsf{ind}}(Z', \mathbf{X})$ and this is a contradiction with the foundation axiom. □

Corollary 3.2.1 *If* $Y \in \mathbf{tc}_{\mathsf{ind}}(Z, \mathbf{X})$, *then* $\mathbf{tc}_{\mathsf{ind}}(Y, \mathbf{X}) \subset \mathbf{tc}_{\mathsf{ind}}(Z, \mathbf{X})$.

PROOF. From Theorem 3.2 we have $\mathbf{tc}_{\mathsf{ind}}(Y, \mathbf{X}) \subseteq \mathbf{tc}_{\mathsf{ind}}(Z, \mathbf{X})$. From Theorem 3.1 (vii) we know that $Y \notin \mathbf{tc}_{\mathsf{ind}}(Y, \mathbf{X})$, so we have $\mathbf{tc}_{\mathsf{ind}}(Y, \mathbf{X}) \subset \mathbf{tc}_{\mathsf{ind}}(Z, \mathbf{X})$. □

In general, an inductive definition induces a corresponding method of proof by induction. The foundation axiom provides the basis from which we can derive a very general and practical inductive methodology.

Theorem 3.3 *Assume p is a general $(k+1)$-ary set predicate such that for fixed sets $\mathbf{X}$ satisfies the following relation for arbitrary values of Z:*

$$\mathsf{ind}(Z, \mathbf{X}) \wedge (q_1(Z, \mathbf{X}) \vee (\forall Y \in \mathbf{tc}_{\mathsf{ind}}(Z, \mathbf{X}))p(Y, \mathbf{X})) \rightarrow p(Z, \mathbf{X}).$$

Then:

(i) *If Z is a set such that* $\mathsf{ind}(Z, \mathbf{X}) \wedge \neg p(Z, \mathbf{X})$, *then there is* $Y \in \mathbf{tc}_{\mathsf{ind}}(Z, \mathbf{X})$ *such that* $\mathsf{ind}(Y, \mathbf{X}) \wedge \neg p(Y, \mathbf{X})$.

(ii) *If Z is any set, then* $\mathsf{ind}(Z, \mathbf{X}) \rightarrow p(Z, \mathbf{X})$.

PROOF. To prove (i) assume that for some set Z we have $\mathsf{ind}(Z, \mathbf{X}) \wedge \neg p(Z, \mathbf{X})$. From $\neg p(Z, \mathbf{X})$ we get $\neg q_1(Z, \mathbf{X})$, and from $\mathsf{ind}(Z, \mathbf{X})$ it follows

$(\exists Y \in \mathbf{tc}_{\mathsf{ind}}(Z, \mathbf{X}))\neg p(Y, \mathbf{X})$. Furthermore, we have $\mathsf{ind}(Y, \mathbf{X})$ by Example 3.2. To prove (ii) we assume a set Z such that $\mathsf{ind}(Z, \mathbf{X}) \wedge \neg p(Z, \mathbf{X})$ and introduce a set V such that

$$V = \{Y \in \mathbf{tc}_{\mathsf{ind}}(Z, \mathbf{X}) :: \neg p(Y, \mathbf{X})\}.$$

By part (i) there is $Y \in V$ such that $\mathsf{ind}(Y, \mathbf{X})$, hence by the foundation axiom there is $Z' \in V$ such that $\mathsf{ind}(Z', \mathbf{X}) \wedge V \cap \mathbf{tc}_{\mathsf{ind}}(Z') = \emptyset$. Again, by part (i) there is $Y' \in \mathbf{tc}_{\mathsf{ind}}(Z', \mathbf{X})$ such that $\neg p(Y', \mathbf{X})$, and by Theorem 3.2 we have $Y' \in \mathbf{tc}_{\mathsf{ind}}(Z, \mathbf{X})$, so $Y' \in V \cap \mathbf{tc}_{\mathsf{ind}}(Z')$, which contradicts the foundation axiom. □

A proof of $\mathsf{ind}(Z, \mathbf{X}) \rightarrow p(Z, \mathbf{X})$ for arbitrary set Z and fixed $\mathbf{X}$ using Theorem 3.3 will be called a proof by ind-*induction on Z over p*, or simply a proof by *induction*. The reference to a general predicate p must be taken literally, even if the given operations and predicates in the rule are basic, in which case the new predicate ind is basic. The predicate p refers to *general predicates*, which are derived by substitution from basic predicates. In fact, many applications of induction in this work involve general (non-basic) predicates.

A proof by ind-induction on Z over p has a rather simple structure. With $\mathbf{X}$ fixed, the object is to prove $\mathsf{ind}(Z, \mathbf{X}) \rightarrow p(Z, \mathbf{X})$ for arbitrary set Z (or, equivalently, that $p(Z, \mathbf{X})$ holds whenever $\mathsf{ind}(Z, \mathbf{X})$ holds). To reach this goal it is sufficient to prove the assumption in Theorem 3.3, hence we assume with Z and $\mathbf{X}$ fixed that $\mathsf{ind}(Z, \mathbf{X}) \wedge \mathrm{IH}$ holds, where IH is an abbreviation for

$$q_1(Z, \mathbf{X}) \vee (\forall Y \in \mathbf{tc}_{\mathsf{ind}}(Z, \mathbf{X}))p(Y, \mathbf{X}),$$

and from this assumption prove $p(Z, \mathbf{X})$ (*the induction step*). This completes the proof of $\mathsf{ind}(Z, \mathbf{X}) \rightarrow p(Z, \mathbf{X})$. We refer to this step by saying that *we prove $p(Z, \mathbf{X})$ by* ind-*induction on Z with $\mathbf{X}$ fixed.* In the induction step the set Z is a fixed arbitrary set that satisfies the assumption $\mathsf{ind}(Z, \mathbf{X}) \wedge$

IH, while the sets $\mathbf{X}$ are assumed fixed, and eventually may satisfy special assumptions.

The assumption $\mathsf{ind}(Z, \mathbf{X}) \wedge \mathrm{IH}$ we call the *induction hypothesis* on Z, although in practice this expression refers mainly to IH, and in a more restricted sense refers to the assumption $(\forall Y \in \mathbf{tc}_{\mathsf{ind}}(Y, \mathbf{X}))p(Y, \mathbf{X})$.

Remark 3.2 In Chapter 10 we define a notion of formal theorem in intuitionistic logic, which is actually a departure from the traditional formulation. Proofs by induction are in fact valid under such logic (see Example 10.2). □

The process of proving $p(Z, \mathbf{X})$ from the induction hypothesis depends on the nature of the predicate p, but still some general rules can be outlined. We note first that the induction hypothesis is a disjunction that induces two obvious cases.

In the first case we assume $\mathsf{ind}(Z, \mathbf{X}) \wedge q_1(Z, \mathbf{X})$ to prove $p(Z, \mathbf{X})$. Very often this case is trivial.

In the second case we assume $\mathsf{ind}(Z, \mathbf{X}) \wedge \neg q_1(Z, \mathbf{X})$, again to prove $p(Z, \mathbf{X})$. In this case, $q_2(Z, \mathbf{X}) \wedge (\forall Y \in \mathbf{tc}_{\mathsf{ind}}(Z, \mathbf{X}))p(Y, \mathbf{X})$ follows from the induction hypothesis. Furthermore, we know $\mathbf{is}_{\mathsf{ind}}(Z, \mathbf{X}) = g(Z, \mathbf{X}) \subseteq \mathbf{tc}_{\mathsf{ind}}(Z, \mathbf{X})$, hence $(\forall Y \in g(Z, \mathbf{X}))p(Y, \mathbf{X})$. In many applications this is the relevant part of the induction hypothesis.

In some special cases the predicate p is such that $\mathbf{tc}_{\mathsf{ind}}(Z, \mathbf{X}) = \emptyset$ implies $p(Z, \mathbf{X})$, so we can assume $\mathbf{tc}_{\mathsf{ind}}(Z, \mathbf{X}) \neq \emptyset$ to prove $p(Z, \mathbf{X})$. This assumption implies $\neg q_1(Z, \mathbf{X})$, so we can move to the second case above and complete the proof from there (see Theorem 3.4).

Note that $\mathsf{ind}(Z, \mathbf{X})$ is always a part of the induction hypothesis, and via this assumption we can use any information available about the predicate ind at the time of the proof. In many applications we do not identify explicitly the predicate p, which is replaced by a boolean term.

Technically, the sets $\mathbf{X}$ are fixed, but if no assumption is made about them, then the proof holds for arbitrary $\mathbf{X}$. If some assumption is made about the sets $\mathbf{X}$, then the proof is valid for any sets $\mathbf{X}$ satisfying the assumption.

EXAMPLE 3.3 Let ind be the inductive predicate in Example 3.1. We want to prove by ind-induction that $\mathsf{ind}(Z) \to \mathsf{ind}(S(Z))$ holds for any set Z. The induction hypothesis is

$$\mathsf{ind}(Z) \wedge (Z = \emptyset \vee (\forall Y \in \mathbf{tc}_{\mathsf{ind}}(Z))\mathsf{ind}(S(Y))).$$

From the induction hypothesis we must prove $\mathsf{ind}(S(Z))$. The case $Z = \emptyset$ is trivial, because $S(\emptyset) = \{\emptyset\}$ and $\mathsf{ind}(\{\emptyset\})$ follows from the induction axiom. If $Z \neq \emptyset$, then from the assumption $\mathsf{ind}(Z)$ and the induction axiom we know that $\emptyset \in Z \wedge (\forall Y \in Z)\mathsf{ind}(Y)$. We conclude that $\emptyset \in S(Z) \wedge (\forall Y \in S(Z))\mathsf{ind}(Y)$. From the induction axiom follows that $\mathsf{ind}(S(Z))$. We have proved that the induction hypothesis on Z implies $\mathsf{ind}(S(Z))$. From Theorem 3.3 we get $\mathsf{ind}(Z) \to \mathsf{ind}(S(Z))$ (note that in this argument $p(Z) \equiv \mathsf{ind}(S(Z))$). Technically, this is a proof by ind-induction. A closer examination shows that the induction hypothesis is not used, and the whole argument depends on the assumption $\mathsf{ind}(Z)$ and the induction axiom. For a real proof by induction, see Theorem 3.4 □

The definition of the predicate ind in Example 3.1 suggests the convenience of using predicate and set terms in inductive definitions to avoid extra definitions that are not essential. So, instead of predicates q_1 and q_2 we may use boolean terms, and instead of the set operations g we may use a set term. Unfortunately, this cannot be done by a simple extension of the basic terms rules, due to the self-referential structure of the induction axiom. On the other hand, note that inductive predicates may occur in basic terms via rule **BT 3**.

Still, we propose to use a more general frame where terms are allowed with the understanding that they can be eliminated by introducing the corresponding predicates and operations. Note that this does not mean we are extending the term formation rules.

Assume U_1 and U_2 are boolean terms and W is a set term, where the free variables are in the list $Z, \mathbf{X}$, and $\mathbf{X}$ is a list of length k. We introduce a $(k+1)$-ary set predicate ind and $(k+1)$-ary set operations $\mathbf{is}_{\mathsf{ind}}$, $\mathbf{tc}_{\mathsf{ind}}$ that satisfy the following axioms:

$$\mathsf{ind}(Z, \mathbf{X}) \equiv (U_1 \vee (U_2 \wedge (\forall Y \in W)\mathsf{ind}(Y, \mathbf{X}))).$$

$$\mathbf{is}_{\mathsf{ind}}(Z, \mathbf{X}) = \{Y \in W :: \neg U_1 \wedge U_2\}.$$

$$\mathbf{tc}_{\mathsf{ind}}(Z, \mathbf{X}) = \mathbf{is}_{\mathsf{ind}}(Z, \mathbf{X}) \cup \bigcup\{\mathbf{tc}_{\mathsf{ind}}(Y, \mathbf{X}) : Y \in \mathbf{is}_{\mathsf{ind}}(Z, \mathbf{X})\}.$$

We consider this type of definition an abbreviation where the predicates q_1, q_2 and the operation g can be derived from the terms U_1, U_2, W. Note that by allowing the use of terms in inductive definitions we are not extending the term rules. Still, an inductive predicate (basic or general) may be a part of a term under rule **BT 3** in Chapter 1.

Remark 3.3 The parameters in a particular application of the induction rule may appear in different forms. For example, consider the induction,

$$\mathsf{ind}(Z, \mathbf{X}) \equiv U_1 \vee (U_2 \wedge (\forall Y \in W)\mathsf{ind}(Y, \mathbf{X})).$$

If X is a parameter that occurs in one of the terms U_1, U_2, W, we say that X is *essential.* Non-essential parameters can be eliminated from the induction if we reduce the arity of the inductive predicate. For example, if ind is a 3-ary inductive predicate of the form,

$$\mathsf{ind}(Z, X, Y') \equiv U_1 \vee (U_2 \wedge (\forall Y \in W)\mathsf{ind}(Y, X, Y')),$$

where Y' is non-essential, then we can introduce a binary predicate ind$'$ in the form,

$$\mathsf{ind}'(Z, X) \equiv U_1 \vee (U_2 \wedge (\forall Y \in W)\mathsf{ind}'(Y, X)).$$

We can reverse this process and, given the predicate ind$'$, we introduce ind by extension with a new parameter Y'. In fact, this happens in some situations where we may need an extra parameter in order to define some recursive operation (see Definition 5.5.1). □

Remark 3.4 The term U_1 can be omitted in applications. This is equivalent to having U_1 a false predicate (say $Z \neq Z$). In this case $\mathsf{ind}(Z, \mathbf{X})$ implies $\mathbf{is}_{\mathsf{ind}}(Z, \mathbf{X}) = \mathrm{W}$, and the closure axiom can be written:

$$\mathbf{tc}_{\mathsf{ind}}(Z) = \mathrm{W} \cup \bigcup\{\mathbf{tc}_{\mathsf{ind}}(Y, \mathbf{X}) : Y \in \mathrm{W} :\}.$$

The term U_2 can also be omitted, so it is equivalent to a true predicate (say $Z = Z$). An induction where U_1 is missing is said to be *deterministic*. An induction where both U_1 and U_2 are missing is said to be *extensional*. This is the simplest form of induction, and plays a substantial role in this work (see Section 3.3). The term W can also be omitted, provided that the variable Y in the body is replaced by a set term V where all the variables are in the list $Z, \mathbf{X}$. In this case the induction takes the form,

$$\mathsf{ind}(Z, \mathbf{X}) \equiv \mathrm{U}_1 \vee (\mathrm{U}_2 \wedge \mathsf{ind}(\mathrm{V}, \mathbf{X})).$$

This is intended to be an abbreviation for the standard induction,

$$\mathsf{ind}(Z, \mathbf{X}) \equiv \mathrm{U}_1 \vee (\mathrm{U}_2 \wedge (\forall Y \in \{\mathrm{V}\})\mathsf{ind}(Y, \mathbf{X})).$$

Note that in this induction we have $g(Z) = \{\mathrm{V}\}$. □

EXERCISE 3.5 Let ind be defined by an extensional induction,

$$\mathsf{ind}(Z) \equiv (\forall Y \in \{Z\})\mathsf{ind}(Y).$$

Let Z be a set. Prove:

(a) $\mathbf{is}_{\mathsf{ind}}(Z) = \{Z\}$.

(b) $\mathsf{ind}(Z) \rightarrow Z \in \mathbf{tc}_{\mathsf{ind}}(Z)$.

(c) $\neg\mathsf{ind}(Z)$.

EXERCISE 3.6 Give an example of a unary inductive predicate ind such that $\mathbf{tc}_{\mathsf{ind}}(Z) = \emptyset \wedge \mathbf{is}_{\mathsf{ind}}(Z) \neq \emptyset$ for every set Z.

EXAMPLE 3.4 Assume q is a given 2-ary general set predicate. We define inductively a 2-ary set predicate ind by the axiom,

$$\mathsf{ind}(Z, X) \equiv (Z = \emptyset \wedge q(\emptyset, X)) \vee \left(Z = \left\{\bigcup Z\right\} \wedge \mathsf{ind}\left(\bigcup Z, \{X\}\right)\right).$$

This is not a legitimate application of the set induction rule, because the parameter X changes to $\{X\}$ in the body of the axiom. A correct application is an axiom of the form,

$$\mathsf{ind}(Z, X) \equiv (Z = \emptyset \wedge q(\emptyset, X)) \vee \left(Z = \left\{\bigcup Z\right\} \wedge \mathsf{ind}\left(\bigcup Z, X\right)\right).$$

In this induction the parameter X is essential. □

EXAMPLE 3.5 Consider an inductive definition given by the axiom,

$$\mathsf{ind}(Z) \equiv Z = \emptyset \vee \left(Z = S\left(\bigcup Z\right) \wedge \mathsf{ind}\left(\bigcup Z\right)\right).$$

This is actually an abbreviation for the induction

$$\mathsf{ind}(Z) \equiv Z = \emptyset \vee \left(Z = S\left(\bigcup Z\right) \wedge \left(\forall Y \in \left\{\bigcup Z\right\}\right) \mathsf{ind}(Y)\right).$$

In this induction we have $\mathbf{is}_{\mathsf{ind}}(Z) = \{\bigcup Z :: Z = S(\bigcup Z)\}$, and $\mathbf{tc}_{\mathsf{ind}}(Z) = [Z = \emptyset \Rightarrow \emptyset, \{\bigcup Z\} \cup \mathbf{tc}_{\mathsf{ind}}(\bigcup Z)]$. □

EXERCISE 3.7 Write an inductive definition of a predicate ind that satisfies the following condition:

$$\mathsf{ind}(Z) \equiv (\forall Y \in g(Z))(\mathsf{ind}(Y) \wedge \mathsf{ind}(g'(Y, Z)),$$

where g and g' are given set operations.

EXAMPLE 3.6 Consider the predicate ind introduced by the induction

$$\mathsf{ind}(Z) \equiv Z = \emptyset \vee \mathsf{ind}\left(\bigcup Z\right).$$

Here we have $\mathbf{is}_{\mathsf{ind}}(Z) = \{\bigcup Z :: Z \neq \emptyset\}$. Another example, although more restricted, is the predicate ind′, introduced by the axiom,

$$\mathsf{ind}'(Z) \equiv Z = \emptyset \vee \left(Z = \left\{\bigcup Z\right\} \wedge \mathsf{ind}'\left(\bigcup Z\right)\right).$$

Here we have $\mathbf{is}_{\mathsf{ind}'}(Z) = \{\bigcup Z :: Z = \{\bigcup Z\}\} = \{Y \in Z :: Z = \{Y\}\}$. □

EXERCISE 3.8 Let ind be the predicate in Example 3.6. Prove:

(a) $\mathsf{ind}(\emptyset)$.

(b) $\mathsf{ind}(\{\{\{\emptyset\}\}\})$.

(c) $\mathsf{ind}(\{\{\{\{\emptyset\}\}\}, \{\emptyset\}\})$.

EXERCISE 3.9 Let ind and ind′ be the predicates of Example 3.6, and Z a set. Prove:

(a) $\mathsf{ind}'(Z) \to \mathsf{ind}(Z)$.

(b) $\mathsf{ind}(Z) \to \mathsf{ind}(\{Z\})$.

(c) $\mathsf{ind}'(Z) \to \mathsf{ind}'(\{Z\})$.

We conclude this section with a useful characterization of the operation $\mathbf{tc}_{\mathsf{ind}}$ that depends on the following predicate, relative to some inductive predicate ind.

3.1.2 $\mathbf{TR}_{\mathsf{ind}}(V, \mathbf{X}) \equiv (\forall Z \in V)\mathbf{is}_{\mathsf{ind}}(Z, \mathbf{X}) \subseteq V$ (V *is* ind-*transitive at* $\mathbf{X}$).

In particular, $\mathbf{TR}_{\mathsf{ind}}(\emptyset, \mathbf{X})$ holds for any inductive predicate ind. Note that $\mathbf{TR}_{\mathsf{ind}}$ depends only on the operation $\mathbf{is}_{\mathsf{ind}}$ and not on the predicate ind.

EXAMPLE 3.7 Let ind be the predicate in Example 3.6. We know that $\mathbf{is}_{\mathsf{ind}}(Z) = \{\bigcup Z :: Z \neq \emptyset\}$. It follows that

$$\mathbf{TR}_{\mathsf{ind}}(V) \equiv (\forall Z \in V)\left(Z = \emptyset \vee \bigcup Z \in V\right).$$

Let $V = \{\{\emptyset\}\}$ and $Z = \{\emptyset\}$. Noting $Z \in V \wedge \mathbf{is}_{\mathsf{ind}}(Z) = Z \wedge \emptyset \notin V$, we conclude that V is not ind-transitive. On the other hand, $Z' = \{\emptyset, \{\emptyset\}\}$ is ind-transitive. □

Theorem 3.4 *Assume* $\mathsf{ind}(Z, \mathbf{X})$ *and* V' *is a set. Then:*

(i) $\mathbf{TR}_{\mathsf{ind}}(\mathbf{tc}_{\mathsf{ind}}(Z, \mathbf{X}), \mathbf{X}) \wedge \mathbf{is}_{\mathsf{ind}}(Z, \mathbf{X}) \subseteq \mathbf{tc}_{\mathsf{ind}}(Z, \mathbf{X})$.

(ii) $\mathbf{TR}_{\mathsf{ind}}(V', \mathbf{X}) \wedge \mathbf{is}_{\mathsf{ind}}(Z, \mathbf{X}) \subseteq V' \rightarrow \mathbf{tc}_{\mathsf{ind}}(Z, \mathbf{X}) \subseteq V'$.

PROOF. Part (i) follows from the transitive axiom and Theorem 3.1 (v). To prove (ii) we use ind-induction on Z over the predicate $p(Z, \mathbf{X}) \equiv \mathbf{is}_{\mathsf{ind}}(Z, \mathbf{X}) \subseteq V' \rightarrow \mathbf{tc}_{\mathsf{ind}}(Z, \mathbf{X}) \subseteq V'$, where V' is fixed and $\mathbf{TR}_{\mathsf{ind}}(V', \mathbf{X})$ is assumed. The induction hypothesis implies $(\forall Y \in \mathbf{tc}_{\mathsf{ind}}(Z, \mathbf{X}))p(Y, \mathbf{X})$. To prove $p(Z, \mathbf{X})$ we assume $\mathbf{is}_{\mathsf{ind}}(Z, \mathbf{X}) \subseteq V' \wedge Y \in \mathbf{tc}_{\mathsf{ind}}(Z, \mathbf{X})$ (to prove $Y \in V'$). There are two cases for Y. The first case is $Y \in \mathbf{is}_{\mathsf{ind}}(Z, \mathbf{X})$, hence $Y \in V'$. The second case is $Y \in \mathbf{tc}_{\mathsf{ind}}(Y', \mathbf{X})$ where $Y' \in \mathbf{is}_{\mathsf{ind}}(Z, \mathbf{X})$, so by the induction hypothesis we have $p(Y', \mathbf{X})$. Since $Y' \in V'$ we have $\mathbf{is}_{\mathsf{ind}}(Y', \mathbf{X}) \subseteq V'$, hence $\mathbf{tc}_{\mathsf{ind}}(Y', \mathbf{X}) \subseteq V'$, and $Y \in V$. □

Remark 3.5 Theorem 3.4 (i) shows that if $V = \mathbf{tc}_{\mathsf{ind}}(Z, \mathbf{X})$ then V is ind-transitive at $\mathbf{X}$ and, furthermore, $\mathbf{is}_{\mathsf{ind}}(Z, \mathbf{X}) \subseteq V$. Part (ii) shows that if V' is another set with the same properties, then $V \subseteq V'$. □

3.2 Discussion

We will consider later more examples of induction, but first we must discuss the primitive construction that supports the rule of induction. We shall show that, given an inductive definition of a predicate ind and operation $\mathbf{tc}_{\mathsf{ind}}$, there is an objective procedure to determine the value of any application of the predicate ind and operation $\mathbf{tc}_{\mathsf{ind}}$ to set arguments in such a way that the four axioms are satisfied. Note that the whole procedure below is described in an informal manner outside the theory. Technically, an inductive predicate enters the system $\mathbb{G}$ via the axioms of the rule.

We restrict the discussion to *standard induction*, where a unary predicate ind is introduced by an axiom of the form,

$$\mathsf{ind}(Z) \equiv q_1(Z) \vee (q_2(Z) \wedge (\forall Y \in g(Z))\mathsf{ind}(Y)).$$

We fix a set Z and describe a process that generates a tree where the root of the tree is the set Z. We call this tree the ind-tree for Z, or simply the *inductive tree for Z*. All nodes in this tree are sets, and the process that generates the tree is controlled by the following rule: if a node Z' has been generated, then the *immediate successors* of Z' are the elements of $\mathbf{is}_{\mathsf{ind}}(Z)$. Note that if $q_1(Z') \vee \neg q_2(Z')$, then there is no immediate successor of Z'. Otherwise, the immediate successors of Z' are the elements of $g(Z')$ (see Definition 3.1.1).

If Z' has no immediate successor (because $\mathbf{is}_{\mathsf{ind}}(Z)$ is empty), we say that Z' is a *halting node*. A *closing node* is a halting node Z' that satisfies the condition, $q_1(Z') \vee q_2(Z')$. We say that a branch in the tree is *closed* if it halts at a closing node Z'.

EXERCISE 3.10 Assume the predicate q_2 is missing in the inductive definition of the predicate ind. Prove that in the inductive tree for any set Z, every halting node is a closing node.

EXERCISE 3.11 Assume that every branch in the inductive tree for Z halts. Prove that no node appears twice in the same branch.

EXERCISE 3.12 Let ind be the inductive predicate in Example 3.1 and $Z = \{\{\emptyset\}\}$. Describe the inductive tree for Z.

We define explicitly the predicate ind as follows: $\mathsf{ind}(Z)$ *holds if and only if every branch in the inductive tree for Z is closed.* If $\mathsf{ind}(Z)$ holds, we define $\mathbf{tc}_{\mathsf{ind}}(Z)$ = the set of all nodes in the inductive tree for Z, excluding the set Z itself. If $\mathsf{ind}(Z)$ does not hold, we set $\mathbf{tc}_{\mathsf{ind}}(Z) = \emptyset$.

Remark 3.6 An important consequence of this definition is that whenever $\mathsf{ind}(Z)$ holds and Z' is a node in the induction tree for Z, then $q_1(Z') \vee q_2(Z')$ holds. If this is not the case, then Z' is a halting node, so there is a branch in the tree which is not closed, and this is a contradiction. □

EXERCISE 3.13 Assume $\mathsf{ind}(Z)$ holds in the sense of the preceding definition. Prove that if Z' is a node in the inductive tree for Z, then $\mathsf{ind}(Z')$ holds.

EXAMPLE 3.8 Assume the predicate ind is introduced by the axiom,

$$\mathsf{ind}(Z) \equiv \mathsf{ind}(\{Z\}).$$

Here $q_1(Z)$ is false, $q_2(Z)$ is true, and $g(Z) = \{\{Z\}\}$. This means that $\mathbf{is}_{\mathsf{ind}}(Z) = \{\{Z\}\}$ and $\{Z\}$ is the only immediate successor of any set Z. We conclude that the inductive tree for Z consists of one branch that is not closed: $Z, \{Z\}, \{\{Z\}\}...$, so $\mathsf{ind}(Z)$ is false for every set Z. □

EXAMPLE 3.9 Assume the predicate ind is introduced by the axiom:

$$\mathsf{ind}(Z) \equiv (\exists Y \in Z) Z = \{Y\} \vee \mathsf{ind}(\{Z\}).$$

Here $q_1(Z) \equiv (\exists Y \in Z) Z = \{Y\}$, $\mathbf{is}_{\mathsf{ind}}(Z) = \{\{Z\} :: (\forall Y \in Z) Z \neq \{Y\}\}$, hence $\mathbf{is}_{\mathsf{ind}}(\{Z\}) = \emptyset$. If $q_1(Z)$ holds, then the induction tree for Z consists of just one closed branch where the only node is Z, hence $\mathsf{ind}(Z)$ holds. If $q_1(Z)$ fails, the tree consists of just one closed branch: $Z, \{Z\}$. Again, this means $\mathsf{ind}(Z)$ holds. We conclude that $\mathsf{ind}(Z)$ holds for every set Z. □

We must prove that the axioms of the induction rule are valid under the above definition of the predicate ind and operation $\mathbf{tc}_{\mathsf{ind}}$. The induction axiom is expressed in the form: $\mathsf{ind}(Z) \equiv \mathsf{ind}^*(Z)$. If $\mathsf{ind}(Z)$ and $q_1(Z)$ it follows that $\mathsf{ind}^*(Z)$. If $\neg q_1(Z)$, then $q_2(Z)$, hence $\mathbf{is}_{\mathsf{ind}}(Z) = g(Z)$, and $(\forall Y \in \mathbf{is}_{\mathsf{ind}}(Z))\mathsf{ind}(Y)$. We conclude $\mathsf{ind}^*(Z)$. In the other direction we assume $\mathsf{ind}^*(Z)$ to prove $\mathsf{ind}(Z)$. If $q_1(Z)$ holds, then Z has no immediate successor and the tree consists of just one branch that closes at Z, so $\mathsf{ind}(Z)$ holds. The situation is similar if $q_2(Z)$ holds and $\mathbf{is}_{\mathsf{ind}}(Z) = \emptyset$. If $\mathbf{is}_{\mathsf{ind}}(Z) \neq \emptyset$ we know that $\mathsf{ind}(Z')$ holds whenever Z' is an immediate successor of Z. This means that all branches of the induction tree for Z are closed, and $\mathsf{ind}(Z)$ holds.

To prove the closure axiom we use extensionality. If $Z' \in \mathbf{tc}_{\mathsf{ind}}(Z)$, then Z' is a node in the inductive tree for Z, different from Z. Clearly, either Z' is an immediate successor of Z, hence $Z' \in \mathbf{is}_{\mathsf{ind}}(Z)$, or there is an immediate successor Y of Z and Z' is a node in the inductive tree for Y. This means $Z' \in \bigcup\{\mathbf{tc}_{\mathsf{ind}}(Y) : Y \in \mathbf{is}_{\mathsf{ind}}(Z) :\}$. In the other direction it is clear that if $Z' \in \mathbf{is}_{\mathsf{ind}}(Z)$, then Z' is a node in the inductive tree for Z, hence $Z' \in \mathbf{tc}_{\mathsf{ind}}(Z)$. Also, if $Z' \in \mathbf{tc}_{\mathsf{ind}}(Y)$ where $Y \in \mathbf{is}_{\mathsf{ind}}(Z)$, then Z' is a node in the inductive tree for Y, hence it is also a node in the inductive tree for Z.

To prove the foundation axiom we consider sets V, Z, such that $Z \in V \wedge \mathsf{ind}(Z) \wedge V \cap \mathbf{tc}_{\mathsf{ind}}(Z) \neq \emptyset$. We generate a branch in the inductive tree for Z, where if Z' is a node in the branch, $Z' \in V$ and $V \cap \mathbf{tc}_{\mathsf{ind}}(Z') \neq \emptyset$, then we take as successor of Z' a node Z'' such that $Z'' \in V$ and $V \cap \mathbf{tc}_{\mathsf{ind}}(Z'') \neq \emptyset$. Since every branch in the tree halts, we eventually reach a node Z' where $Z' \in V \wedge V \cap \mathbf{tc}_{\mathsf{ind}}(Z') \neq \emptyset$, and no successor Z'' of Z' satisfies the above conditions. Under these assumptions there is at least one immediate successor Z'' of Z' such that $Z'' \in V \cap \mathbf{tc}_{\mathsf{ind}}(Z')$, and this Z'' satisfies the condition in the foundation axiom.

The validity of the transitive axiom is immediate, for if Z' is a node in the tree for Z, then all the immediate successors of Z' are nodes in the tree, different from Z.

The notion of tree in the above construction is intended as a *process* that may continue indefinitely, and not as a complete totality. The requirement in the definition of $\mathsf{ind}(Z)$ in the sense that every branch of the inductive tree is closed is not intended as a reference to to a set of branches, but rather to a structural property of the tree as a process. On the other hand, we shall later introduce constructions where the totality of all branches of a tree is a set, and quantification over such totality is legitimate (see Section A.4 in the Appendix). Still, such constructions involve the inductive and recursive properties of the set ω (see Chapter 7), so some amount of induction via primitive constructions seems to be unavoidable. Furthermore, the recursion rule, as introduced in Chapter 5, cannot be derived from the explicit definition, as the standard inductive proof (for transfinite recursion) involves a non-local predicate.

Rather than pushing an artificial derivation where induction is the result of a higher order construction, depending on the set ω and some enumeration rule, we take the position that induction, as explained above in terms of induction trees, is a fundamental mathematical principle. There is no doubt that to some extent the principle of induction is being abused in mathematical practice, and reduced to a verbal iteration procedure. Still, the theory we propose in this section is extremely conservative, providing an environment in which the validity of the induction axioms is completely transparent.

3.3 Extensional Induction

Let p be a general binary predicate. In some situations we may want to consider p as a form of ordering, and when $p(Y, Z)$ holds we say that Y *p-precedes* Z. A set Z is p-well-founded if whenever V is a set such that $Z \in V$, then there is a set $Z' \in V$ and $(\forall Y \in V)\neg p(Y, Z')$. Informally, this means that if we attempt to generate a sequence $Z_0, Z_1, ..., Z_n, ...$ where $Z_0 = Z$ and for every n $Z_n \in V$ and Z_{n+1} p-precedes Z_n, eventually we reach an element Z_n where it is impossible to define Z_{n+1}. We would like to characterize by set induction the sets that are p-well-founded, and so be able to treat p as a kind of well-ordering. In particular, we should be able to define operations by recursion on the ordering p.

EXAMPLE 3.10 For example, let $p(Y, Z) \equiv Y \in Z \wedge \emptyset \in Z$, so $p(Y, Z) \equiv Y \in \mathbf{is}_{\mathsf{ind}}(Z)$ where ind is the predicate in Example 3.1. Clearly, $\emptyset$ is p-well-founded, and in fact every set Z such that $\emptyset \notin Z$ is p-well-founded. Note that if $\emptyset \notin Z$ and $Z \neq \emptyset$, then $\mathsf{ind}(Z)$ fails. On the other hand, if $\mathsf{ind}(Z)$ holds, then by the foundation axiom it follows that Z is p-well-founded. We conclude that the predicate ind does not provide a correct characterization of the p-well-founded sets. □

Remark 3.7 In general, we can take an inductive predicate ind and define $p(Y, Z) \equiv Y \in \mathbf{is}_{\mathsf{ind}}(Z)$, and from the foundation axiom it follows that $\mathsf{ind}(Z)$ implies that Z is p-well-founded. The preceding example shows that the converse of this relation is not necessarily true. This means we must impose restrictions on the induction to make sure that we obtain a correct characterization of the sets that are p-well-founded. □

An adequate theory can be derived by assuming that the induction is extensional, in the sense of Remark 3.2 (see also Remark 3.8). So, now ind is $(k+2)$-ary predicate defined inductively in the form,

$$\mathsf{ind}(Z, \mathbf{X}) \equiv (\forall Y \in g(Z, \mathbf{X}))\mathsf{ind}(Z, \mathbf{X}),$$

where we assume parameters $\mathbf{X}$ to obtain a more general theory. This induction is completely determined by the set operation g, and the predicate p is defined by $p(Y, Z, \mathbf{X}) \equiv Y \in g(Z, \mathbf{X})$. This relation indicates that p is an extensional predicate, in the sense of Section 2.2, and in fact every extensional predicate can be generated in this way. Recall that in Section 2.2 we describe a number of constructions that can be used to generate extensional predicates.

Since the predicate p and the set operation g are determined one by the other, it is convenient to leave out the predicate and work directly with the operation g. With this understanding, if g is a given $(k+1)$-ary predicate we define by induction the predicate:

3.3.1 $\mathbf{WF}^*_g(Z,\mathbf{X}) \equiv (\forall Y \in g(Z,\mathbf{X}))\mathbf{WF}^*_g(Y,\mathbf{X})$. (*$Z$ is g-well-founded at $\mathbf{X}$*)

We say that the predicate $\mathbf{WF}^*_g$ is *extensionally induced by the set operation g*.

The corresponding transitive closure operation is denoted by $\mathbf{tc}^*_g$. Noting that $\mathbf{is}_{\mathbf{WF}^*_g}(Z,\mathbf{X}) = g(Z,\mathbf{X})$, the closure axiom can be written in the form,

$$\mathbf{tc}^*_g(Z,\mathbf{X}) = g(Z,\mathbf{X}) \cup \bigcup\{\mathbf{tc}^*_g(Y,\mathbf{X}) : Y \in g(Z,\mathbf{X}) :\}.$$

The notation for the predicate $\mathbf{TR}_{\mathsf{ind}}$ also changes and we write $\mathbf{TR}^*_g(V,\mathbf{X}) \equiv \mathbf{TR}_{\mathbf{WF}^*_g}(Z,\mathbf{X})$. The formal definition takes the form,

$$\mathbf{TR}^*_g(V,\mathbf{X}) \equiv (\forall Z \in V)g(Z,\mathbf{X}) \subseteq V.$$

When the induction is extensional, proofs by induction can be arranged with the induction hypothesis reduced to a deterministic assumption. In fact, in proving $p(Z,\mathbf{X})$ the induction hypothesis is simply $\mathbf{WF}^*_g(Z,\mathbf{X}) \wedge (\forall Y \in \mathbf{tc}^*_g(Z,\mathbf{X}))p(Y,\mathbf{X})$. In particular, since $g(Z,\mathbf{X}) \subseteq \mathbf{tc}^*_g(Z,\mathbf{X})$, we have $(\forall Y \in g(Z,\mathbf{X}))p(Y,\mathbf{X})$.

Remark 3.8 Given a $(k+2)$-ary operation g, we fix $\mathbf{X}$ and introduce the predicate $p(Y,Z) \equiv Y \in g(Z,\mathbf{X})$. From the foundation axiom it follows that whenever Z is g-well-founded at $\mathbf{X}$, then Z is p-well-founded. From Corollary A.22.2 it follows that whenever Z is not g-well-founded at $\mathbf{X}$ there is a set V such that $Z \in V$ and $(\forall Z' \in V)V \cap g(Z') \neq \emptyset$. Hence, if p is a given extensional predicate with extensional operation g in the sense of Section 2.2, then $\mathbf{WF}^*_g$ characterizes the sets that are p-well-founded. □

Remark 3.9 We can prove that any induction where the predicate q_2 does not occur can be reduced to an extensional induction. Let ind be a predicate introduced by an induction of the form,

$$\mathsf{ind}(Z) \equiv q_1(Z) \vee (\forall Y \in g(Z))\mathsf{ind}(Y).$$

We can replace this induction with the following extensional induction,

$$\mathsf{ind}'(Z) \equiv (\forall Y \in \mathbf{is}_{\mathsf{ind}}(Z))\mathsf{ind}'(Y).$$

Note that the set operation $\mathbf{is}_{\mathsf{ind}}$ in the second induction is only notationally related to the inductive predicate ind (see Definition 3.1.1). □

EXERCISE 3.14 Let ind and ind′ be the predicates in Remark 3.9. Prove $\mathsf{ind}(Z) \equiv \mathsf{ind}'(Z)$ for any set Z.

Theorem 3.5 *Assume the set Z is g-well-founded at $\mathbf{X}$. Then:*

(i) $Z \notin \mathbf{tc}_g^*(Z, \mathbf{X})$.

(ii) *If $Y \in \mathbf{tc}_g^*(Z, \mathbf{X})$, then $Z \notin \mathbf{tc}_g^*(Y, \mathbf{X})$.*

(iii) *If $Y \in \mathbf{tc}_g^*(Z, \mathbf{X})$ and $W \in \mathbf{tc}_g^*(Y, \mathbf{X})$, then $Z \notin \mathbf{tc}_g^*(W, \mathbf{X})$.*

PROOF. Part (i) follows from Theorem 3.1 (vii). Part (ii) follows from the foundation axiom with $V = \{Y, Z\}$. Part (iii) follows from the foundation axiom with $V = \{W, Y, Z\}$. □

An important example of extensional induction is obtained by taking $g(Z) = Z$. In this case we omit the subscript and the superscript.

3.3.2 $\mathbf{WF}(Z) \equiv (\forall Y \in Z)\mathbf{WF}(Y)$ (*Z is well-founded*).

3.3.3 $\mathrm{tc}(Z) = \mathbf{tc}_{\mathbf{WF}}(Z)$ (*the transitive closure of Z*).

Note that $\mathbf{is}_{\mathbf{WF}}(Z) = g(Z) = Z$ and whenever Z is well-founded and $Y \in Z$, then Y is well-founded. The predicate p induced by this extensional induction is $p(Y, Z) \equiv Y \in Z$, so we are considering the predicate $\in$ as an ordering where Y p-precedes Z if and only if $Y \in Z$.

Remark 3.10 The predicate **WF** is universal in the sense that whenever f is a basic operation in the system $\mathbb{G}$ and the sets $\mathbf{X}$ are well-founded, then $f(\mathbf{X})$ is well-founded. Note that we leave open the possibility of extending the system $\mathbb{G}$ with new rules, in which case the universality of **WF** may not be preserved. □

EXERCISE 3.15 Let V be a set such that $(\exists Z \in V)\mathbf{WF}(Z)$. Prove that $(\exists Z \in V)(\mathbf{WF}(Z) \wedge \mathbf{tc}(Z) \cap V = \emptyset)$.

EXERCISE 3.16 Assume Z is well-founded and $Y \subseteq Z$. Prove that Y is well-founded.

EXERCISE 3.17 Assume Z is well-founded. Describe the process that generates the **WF**-inductive tree for Z.

Theorem 3.6 *Let Z be a well-founded set. Then:*

(i) $Z \subseteq \mathbf{tc}(Z)$

(ii) $(\forall Y \in \mathbf{tc}(Z))\mathbf{tc}(Y) \subseteq \mathbf{tc}(Z)$.

(iii) $\mathbf{TR}(Z) \equiv \mathbf{TR}_{\mathbf{WF}}(Z)$.

(iv) $\mathbf{tc}(Z)$ *is transitive.*

(v) *If V is a transitive set such that $Z \subseteq V$, then $\mathbf{tc}(Z) \subseteq V$.*

(vi) $\mathbf{TR}(Z) \equiv \mathbf{tc}(Z) = Z$.

PROOF. Parts (i) and (ii) follow from Theorem 3.1 (v) and (vi). In part (iii) the implication $\mathbf{TR}(Z) \to \mathbf{TR}_{\mathbf{WF}}(Z)$ is clear from Definition 1.6.7 (even if Z is not well-founded). In the other direction we note that if $\mathbf{TR}_{\mathbf{WF}}(Z) \wedge Y \in Z$, then $\mathbf{WF}(Y)$, so $Y \subseteq Z$. Part (iv) comes from Corollary 3.4.1 (i) and part (iii). Part (v) comes from Theorem 3.4 (ii), noting $\mathbf{TR}_{\mathbf{WF}}(V)$. In part (vi) the implication from right to left comes from Theorem 3.4 (i) and part (iii). In the other direction we know that $Z \subseteq \mathbf{tc}(Z)$, and the inclusion $\mathbf{tc}(Z) \subseteq Z$ follows from part (v) with $V = Z$. □

EXAMPLE 3.11 Let g be the unary set operation such that $g(Z) = \{Y \in Z :: Y \neq Z\}$. It follows that if Z is a well-founded set, then $g(Z) = Z$. From this we can prove $\mathbf{WF}^*_g(Z)$ if Z is well-founded. On the other hand, if A is a set such that $A = \{A\}$, then A is not well-founded and $g(A) = \emptyset$, so $\mathbf{WF}^*_g(A)$ holds and A is g-well-founded. □

EXERCISE 3.18 Let g be the set operation in Example 3.11. Prove that $\mathbf{WF}(Z) \to \mathbf{WF}^*_g(Z)$.

EXERCISE 3.19 Let ind be the predicate in Example 3.1 and Z a set such that $\mathsf{ind}(Z)$ holds. Prove:

(a) $\mathbf{WF}(Z)$.

(b) $\mathbf{tc}_{\mathsf{ind}}(Z) = \mathbf{tc}(Z)$.

EXAMPLE 3.12 Another example is the extensional induction $\mathbf{WF}^*_g$ induced by the set operation $g(Z, X) = Z \cap X$. This means that

$$\mathbf{WF}^*_g(Z, X) \equiv (\forall Y \in Z \cap X)\mathbf{WF}^*_g(Y, X).$$

This predicate may be considered a relativization of the predicate $\mathbf{WF}$ to a parameter X and it is closely related with the recursive operation collapsing introduced in Section 5.5 (see Definition 5.5.1). □

EXERCISE 3.20 Let g be the set operation in Example 3.12 and X, Z sets. Prove:

(a) $\mathbf{WF}(Z) \to \mathbf{WF}^*_g(Z, X)$.

(b) $\mathbf{WF}(X) \to \mathbf{WF}^*_g(Z, X)$.

EXERCISE 3.21 Assume Z and X are sets, Z is well-founded and $S(Z) = S(X)$. Prove $Z = X$.

EXERCISE 3.22 Assume Z is well-founded and X is transitive. Prove that the following conditions are equivalent:

(a) $X \subset S(Z)$.

(b) $X \subseteq Z$.

EXERCISE 3.23 Assume the set Z is well-founded. Prove:

(a) $S(Z) \setminus Z = \{Z\}$.

(b) $S(Z) \ominus Z = Z$.

(c) If $S(Z)$ is transitive, then Z is transitive.

EXERCISE 3.24 Let f a unary set operation, introduced by replacement in the form

$$f(X) = \{Y \in X :: \mathbf{WF}(Y)\}.$$

Prove that $f(X)$ is well-founded and $f(X) \notin X$.

3.4 Notes

General rules of induction and recursion are also available in standard set theory (see Tarski (1955), Montague (1955) and Levy (1979)). We take set induction (and set recursion in Chapter 5) as primitive rules. The underlying primitive construction given in the text is crucial, and the formal axioms are derived from the construction.

The rule of set induction is similar, but not identical, to local induction (which we consider explicitly in Chapters 7 and 8). Local induction is a general mathematical procedure where the purpose is to introduce a set satisfying some closure properties. In applications the set is usually a relation or a function. It is well-known (and we show this in Chapter 8) that the formalization of this procedure involves higher set theory and, eventually, the power-set construction.

We have chosen a particularly restrictive form of induction, that can be called non-deterministic (see Remark 3.4). A more general type of non-deterministic induction can be obtained by allowing existential quantification. For example, an inductive definition of the predicate **FI** can be given in the form $\mathbf{FI}(Z) \equiv Z = \emptyset \vee (\exists Y \in Z)\mathbf{FI}(Z \ominus Y)$, with a rule of proof by induction derived from this condition. This type of induction presents a number of problems, and it is not included in this work.

On the other hand, there are inductions that are deterministic, but are outside the frame defined by the rule of set induction. Eventually, they will have to be incorporated into the scope of the induction rule (see Exercise 3.7).

We define well-foundedness as a property that is not necessarily universal (see Remark 3.10), and in this way avoid the standard foundation axiom (not to be confused with the foundation axiom in the induction rule in Section 3.1). This approach goes back to Mirimanoff (1917) (for details see Levy (1979)). An interesting consequence is that, in principle, the theory allows for construction rules of non-well-founded sets in the sense of Aczel (1988). In fact, the graph constructions presented in Aczel's monograph appear to be primitive constructions in the sense we are using here. The proper formal axiom would be the equations derived from the graphs.

Extensional induction provides the basis for the theory of well-founded relations introduced in Chapter 4.

Chapter 4

Applications

In this section we define and study a number of classical constructions in set theory. The main tool here is the general rule of induction in Chapter 3, with the associated rule of proof by induction provided by Theorem 3.3. With every inductive predicate we write also the definitions of the operations $\mathbf{is}_{\mathsf{ind}}$ and $\mathbf{tc}_{\mathsf{ind}}$.

In the next chapter we discuss applications where set operations are introduced by recursion supported by the inductive predicates defined below.

4.1 Finite Sets

The first application is the unary predicate **FI**, which is introduced by the following induction.

4.1.1 $\mathbf{FI}(Z) \equiv (\forall V \in \{Z \ominus Y : Y \in Z :\})\mathbf{FI}(V)$ (*Z is finite*).

4.1.2 $\mathbf{is}_{\mathbf{FI}}(Z) = \{Z \ominus Y : Y \in Z :\}$.

4.1.3 $\mathbf{tc}_{\mathbf{FI}}(Z) = \mathbf{is}_{\mathbf{FI}}(Z) \cup \bigcup\{\mathbf{tc}_{\mathbf{FI}}(Z \ominus Y) : Y \in Z :\}$.

Note that from the induction axiom it follows that the set $\emptyset$ is finite. Furthermore, when using **FI**-induction over a predicate p the induction hypothesis is $p(X)$ for every set $X \in \mathbf{tc_{FI}}(Z)$. In most applications we need only the assumption $p(Z \ominus Y)$ for $Y \in Z$.

EXERCISE 4.1 Let X and Y be arbitrary sets. Prove:

(a) $\{X, Y\}$ is a finite sets.

(b) $\mathbf{tc_{FI}}(\{X, Y\}) = \{\emptyset, \{X\}, \{Y\}\}$.

Theorem 4.1 *Let X be an arbitrary set and Z a finite set. Then:*

(i) *$\{X\}$ is a finite set.*

(ii) *$Z \cap X$ is a finite set.*

(iii) *$Z; X$ is a finite set.*

(iv) *If $X \in \mathbf{tc_{FI}}(Z)$, then $(\exists V \in Z) X \subseteq Z \ominus V$.*

PROOF. To prove (i), note that if $Y \in \{X\}$, then $Y = X$ and $\{X\} \ominus X = \emptyset$, which is finite. It follows that $\{X\}$ is finite. To prove (ii) we use **FI**-induction with X fixed over the predicate $p(Z) \equiv \mathbf{FI}(Z \cap X)$ and prove that $\mathbf{FI}(Z) \to p(Z)$ (note that p is a general set predicate rather than a basic set predicate). The induction hypothesis is that Z is finite and whenever $Y \in Z$, then $(Z \ominus Y) \cap X$ is finite. Since $(Z \cap X) \ominus Y = (Z \ominus Y) \cap X$, it follows that $Z \cap X$ is finite. In part (iii) with X fixed we define $p'(Z) \equiv \mathbf{FI}(Z; X)$ and prove by **FI**-induction that $\mathbf{FI}(Z) \to p'(Z)$. The induction hypothesis is that Z is finite, and whenever $Y \in Z$, then $(Z \ominus Y); X$ is finite. To prove that $Z; X$ is finite we may assume that $X \notin Z$. Hence if $Y = X$ we have $(Z; X) \ominus Y = Z$ is finite, and if $Y \neq X$, then $(Z; X) \ominus Y = (Z \ominus Y); X$ is also finite. It follows that $Z; X$ is finite. Part (iv) follows by **FI**-induction with X fixed. If $X \in \mathbf{tc_{FI}}(Z)$, then by the closure axiom there is $V \in Z$ such that either $X = Z \ominus V$ or $X \in \mathbf{tc_{FI}}(Z \ominus X)$. By the induction hypothesis there is $V' \in Z \ominus V$ such that $X \subseteq (Z \ominus V) \ominus V' \subseteq Z \ominus V$. □

Corollary 4.1.1 *If Z is finite and $X \subseteq Z$, then X is finite.*

PROOF. Immediate from Theorem 4.1 (ii), since in this case $Z \cap X = X$.□

Corollary 4.1.2 *If there is $Y \in Z$ such that $Z \ominus Y$ is finite, then Z is finite.*

PROOF. Immediate from Theorem 4.1 (iii), noting that in this case $Z = (Z \ominus Y); Y$. □

Corollary 4.1.3 *If Z is a finite set, then $S(Z)$ is also a finite set.*

PROOF. Immediate from Theorem 4.1 (iii). □

Corollary 4.1.4 *Let F be a function and Z a finite set. Then $F[Z]$ is finite.*

PROOF. The proof is **FI**-induction on Z. If $Z = \emptyset$, then $F[Z] = \emptyset$. Otherwise there is $Y \in Z$ and $F[Z \ominus Y]$ is finite by the induction hypothesis. Since $F[Z] = F[Z \ominus Y]; \mathbf{ap}(F, Y)$ it follows that $F[Z]$ is finite. □

EXERCISE 4.2 Assume F is a function. Prove:

(a) If $\mathbf{do}(F)$ is finite, then $\mathbf{ra}(F)$ is finite.

(b) If $\mathbf{ra}(F)$ is finite and F is 1–1, then $\mathbf{do}(F)$ is finite.

Remark 4.1 The fact that the image of a finite set under a function is a finite set (Corollary 4.1.4) is often applied in a syntactical context. For example, assume a set X is given by replacement in the form,

$$X = \{f(Y) : Y \in W : p(Y) :\},$$

where W is a finite set, f is an operation, and p is a predicate. We can describe this set by means of the function $F = (\lambda Y \in W)f(Y)$, where $\mathbf{do}(F) = W$, so $X = F[Z]$, and $Z = \{Y \in W :: p(Y)\}$ is a subset of W. We conclude that Z is finite, hence X is also finite. If we know that $f(Y)$ is finite whenever $Y \in W$ and $p(Y)$ holds, we may conclude that $\bigcup X$ is a finite set (see Corollary 4.3.1 below). □

EXERCISE 4.3 Prove that if Z is finite, then $\mathbf{is}_{\mathbf{FI}}(Z)$ is also finite.

Theorem 4.2 *If F is a net and Z is a finite set such that $Z \subseteq \bigcup \mathbf{ra}(F)$, then there is $V \in \mathbf{do}(F)$ such that $Z \subseteq \mathbf{ap}(F, V)$.*

PROOF. The proof is by **FI**-induction on Z. The case $Z = \emptyset$ is trivial, so we may assume there is $Y \in Z$, and by the induction hypothesis there is $V' \in \mathbf{do}(F)$ such that $Z \ominus Y \subseteq \mathbf{ap}(F, V')$. Also there is $V'' \in \mathbf{do}(F)$ such that $Y \in \mathbf{ap}(F, V'')$. By Definition 1.6.8 there is $V \in \mathbf{do}(F)$ such that $V' \subseteq V$ and $V'' \subseteq V$. By Definition 2.3.12 it follows that $\mathbf{ap}(F, V') \subseteq \mathbf{ap}(F, V)$ and $\mathbf{ap}(F, V'') \subseteq \mathbf{ap}(F, V)$. We conclude that $Z \subseteq \mathbf{ap}(F, V)$. □

EXERCISE 4.4 Assume Z is a finite set and F is a 1–1 function such that $F[Z] \subseteq Z$. Prove that $F[Z] = Z$.

EXERCISE 4.5 Assume Z is a finite set and F a 1–1 function such that $Z \subseteq F[Z]$. Prove that $F[Z] = Z$.

Theorem 4.3 *If Z and X are finite sets, then $Z \cup X$ is a finite set.*

PROOF. By **FI**-induction on Z with X fixed. If $Z = \emptyset$ we have $Z \cup X = X$, which is finite. Otherwise, there is $Y \in Z$, and we can write

$$Z \cup X = ((Z \ominus Y) \cup X); Y,$$

so the conclusion follows immediately from the induction hypothesis and Theorem 4.1 (iii). □

Corollary 4.3.1 *If Z is a finite set and the elements of Z are finite sets, then $\bigcup Z$ is a finite set.*

PROOF. By **FI**-induction on Z. The induction hypothesis is that Z is finite and $\bigcup(Z \ominus Y)$ is finite whenever $Y \in Z$. If Z is empty we have $\bigcup Z = Z$. Otherwise, there is $Y \in Z$, and we can write $\bigcup Z = \bigcup(Z \ominus Y) \cup Y$, hence $\bigcup Z$ is finite by the induction hypothesis and Theorem 4.3. □

EXERCISE 4.6 Prove that if $\bigcup Z$ is finite and $X \in Z$, then X is finite.

Corollary 4.3.2 *If Z is a finite set, then $\mathbf{tc_{FI}}(Z)$ is also a finite set.*

PROOF. The proof is by **FI**-induction on Z. From the induction hypothesis it follows that $\mathbf{tc_{FI}}(Z)$ is the union of a finite set where the elements are finite, so it is finite by Corollary 4.3.1 (see Remark 4.1). □

Theorem 4.4 *Let Z and X be sets and assume Z is finite. Then:*

(i) *$\{X\} \times Z$ is finite.*

(ii) *If X is finite, then $X \times Z$ is finite.*

PROOF. Part (i) follows from Corollary 4.1.4 and the function $F = \{<Y, <X, Y>> : Y \in Z :\}$. In part (ii) we note that $X \times Z = \bigcup\{\{Y\} \times Z : Y \in X\}$, so it is finite by Theorem 4.3 (see Remark 4.1). □

Corollary 4.4.1 *If $Z \times X$ is finite and $Z \times X \neq \emptyset$, then both Z and X are finite.*

PROOF. Since $Z \times X$ is finite and contains only finite sets, then $\bigcup(Z \times X)$ is finite. Again, the elements of $\bigcup(Z \times X)$ are finite, so it follows that $\bigcup\bigcup(Z \times X)$ is finite. Furthermore, $Z \cup X = \bigcup\bigcup(Z \times X)$, so Z and X are finite. □

Theorem 4.5 *Let Z be a non-empty finite set. There is $Y \in Z$, which is maximal in the sense that if $Y' \in Z$ and $Y \subseteq Y'$, then $Y = Y'$.*

PROOF. We use **FI**-induction. The induction hypothesis is that Z is finite and non-empty, and whenever $Y \in Z$ and $Z \ominus Y$ is non-empty, then there is $Y' \in Z \ominus Y$ and Y' is maximal in $Z \ominus Y$. We fix $Y \in Z$ and note that if $Z \ominus Y$ is empty, then Y is maximal in Z. Otherwise, there is Y' which is maximal in $Z \ominus Y$. If $Y' \subseteq Y$ it follows that Y is maximal in Z. If $Y' \not\subseteq Y$, then Y' is maximal in Z. □

EXERCISE 4.7 Let Z be a non-empty finite directed set. Prove that there is $Y \in Z$, which is maximal in the sense that if $Y' \in Z$, then $Y' \subseteq Y$.

EXERCISE 4.8 Assume the set Z is finite and non-empty. Prove that there is a set $X \in Z$ that is *minimal*, in the sense that whenever $Y \in Z$ and $Y \subseteq X$, then $Y = X$.

Theorem 4.6 *Let Z be a finite set and X an arbitrary set. The following conditions are equivalent:*

(i) $X \subset Z$.

(ii) $X \in \mathbf{tc_{FI}}(Z)$.

PROOF. The implication from (i) to (ii) follows by **FI**-induction on Z with X fixed. The predicate p is defined as

$$p(Z) \equiv X \subset Z \to X \in \mathbf{tc_{FI}}(Z).$$

From the induction hypothesis we know that Z is finite and $p(Z \ominus Y)$ holds whenever $Y \in Z$. We assume $X \subset Z$, so there is $Y \in Z$ such that $X \subseteq Z \ominus Y$. If $X = Z \ominus Y$, then $X \in \mathbf{is_{FI}}(Z) \subseteq \mathbf{tc_{FI}}(Z)$. Otherwise $X \subset Z \ominus Y$, so by the induction hypothesis we have $X \in \mathbf{tc_{FI}}(Z \ominus Y) \subseteq \mathbf{tc_{FI}}(Z)$. The implication from (ii) to (i) is also by **FI**-induction. If $X \in \mathbf{tc_{FI}}(Z)$, then there is $Y \in Z$ such that either $X = Z \ominus Y$, so $X \subset Z$, or $X \in \mathbf{tc_{FI}}(Z \ominus Y)$, so by the induction hypothesis we have $X \subset Z \ominus Y \subset Z$. □

Corollary 4.6.1 *Assume $Z' \subseteq Z$, where Z is a finite set. Then,* $\mathbf{tc_{FI}}(Z') \subseteq \mathbf{tc_{FI}}(Z)$.

PROOF. If $X \in \mathbf{tc_{FI}}(Z')$, then $X \subset Z' \subseteq Z$, so by Theorem 4.6 we have $X \in \mathbf{tc_{FI}}(Z)$. □

Corollary 4.6.2 *Assume Z is finite and Y is an arbitrary set. The following conditions are equivalent:*

(i) $Y \in \mathbf{tc_{FI}}(Z)$.

(ii) $(\exists V \in Z) Y \subseteq Z \ominus V$.

PROOF. Immediate from Theorem 4.6. □

Remark 4.2 From Theorem 4.6 it follows that whenever Z is a finite set then $\mathbf{tc_{FI}}(Z) \cup \{Z\}$ is the set of all subsets of Z, usually called the power-set of Z. This is an important construction, but later we define a more general operation that, given an arbitrary set Z, generates the set of all finite subsets of Z (see Definition 5.4.2). □

We include in this section another inductive predicate derived from the predicate **FI**.

4.1.4 $\mathbf{HF}(Z) \equiv \mathbf{FI}(Z) \wedge (\forall Y \in Z)\mathbf{HF}(Y)$ (*Z is hereditarily finite*).

4.1.5 $\mathbf{is_{HF}}(Z) = [\mathbf{FI}(Z) \Rrightarrow Z, \emptyset]$.

4.1.6 $\mathbf{tc_{HF}}(Z) = [\mathbf{FI}(Z) \Rrightarrow Z \cup \bigcup\{\mathbf{tc_{HF}}(Y) : Y \in Z :\}, \emptyset]$.

Note that the induction 4.1.4 is not extensional.

Theorem 4.7 *Let Z be hereditarily finite and Y a set. Then:*

(i) *If $Y \subseteq Z$, then Y is hereditarily finite.*

(ii) *$Z \setminus Y$ is hereditarily finite.*

(iii) *$\bigcap Z$ is hereditarily finite.*

(iv) *$\bigcup Z$ is hereditarily finite.*

(v) *If Y is hereditarily finite, then $Z \cup Y$ is hereditarily finite.*

(vi) *$[Z]_1$ and $[Z]_2$ are hereditarily finite.*

PROOF. To prove (i), note that Y is finite and the elements of Y are hereditarily finite. Part (ii) follows from (i), noting that $Z \setminus Y \subseteq Z$. To prove (iii), note that either $Z = \emptyset$ or there is $Z' \in Z$ where Z' is hereditarily finite and $\bigcap Z \subseteq Z'$. Part (iv) follows from Corollary 4.3.1 and Definition 4.1.4. Part (v) follows from (iv), noting that $\{Z, Y\}$ is hereditarily finite. Part (vi) follows from Definitions 2.3.1 and 2.3.2. □

Corollary 4.7.1 *If X and Y be hereditarily finite sets, then:*

(i) $<X, Y>$ *is hereditarily finite.*

(ii) $X \times Y$ *is hereditarily finite.*

PROOF. Part (i) is clear, noting that $<X, Y>$ is generated by finite unions with hereditarily finite sets. Part (ii) follows from part (i) and Theorem 4.4 (ii). □

EXERCISE 4.9 Prove that if Z is hereditarily finite, then $S(Z)$ is also hereditarily finite.

EXERCISE 4.10 Assume U is a set such that $\emptyset \in U$, and furthermore U is closed under the insertion operation (so whenever X and Y are elements of U, then $X;Y$ is also an element of U). Prove:

(a) If a set X is finite and every element of X is in U, then $X \in U$.

(b) If a set X is hereditarily finite, then $X \in U$.

EXERCISE 4.11 Assume Z is a set. Prove:

(a) If Z is hereditarily finite, then $\mathbf{tc_{FI}}(Z)$ is also hereditarily finite.

(b) If $\bigcup Z$ is finite, then $Z \subseteq \mathbf{tc_{FI}}(\bigcup Z) \cup \{\bigcup Z\}$.

(c) If $\bigcup Z$ is finite, then Z is finite.

(d) If $\bigcup Z$ is hereditarily finite, then Z is hereditarily finite.

The definition of the hereditarily finite sets can be generalized as follows. Assume p is a general unary set predicate and define by induction the following predicate $p^\dagger$:

4.1.7 $p^\dagger(Z) \equiv p(Z) \wedge (\forall Y \in Z)p^\dagger(Y)$ (*Z is an hereditarily p-set*).

4.1.8 $\mathbf{is}_{p^\dagger}(Z) = [p(Z) \Rrightarrow Z, \emptyset]$.

4.1.9 $\mathbf{tc}_{p^\dagger}(Z) = [p(Z) \Rrightarrow Z \cup \bigcup\{\mathbf{tc}_{p^\dagger}(Y) : Y \in Z :\}, \emptyset]$.

Remark 4.3 The properties in Theorem 4.7 cannot be extended to the hereditarily p-sets, because they depend on special properties of the finite sets. Below we prove several general properties of the hereditarily p-set that are valid, in particular, for the hereditarily finite sets. □

Theorem 4.8 *Let p be a general set predicate and Z be an arbitrary set. Then:*

(i) $p^\dagger(Z) \to (\forall Y \in \mathbf{tc}_{p^\dagger}(Z))p(Y)$.

(ii $p^\dagger(Z) \to \mathbf{WF}(Z) \wedge \mathbf{tc}(Z) = \mathbf{tc}_{p^\dagger}(Z)$.

(iii) $p^\dagger(Z) \to (\forall Y \in \mathbf{tc}(Z))p(Y)$.

PROOF. The three properties follow by $p^\dagger$-induction. In part (i) the induction hypothesis is $p^\dagger(Z) \wedge (\forall Y \in Z)(\forall Y' \in \mathbf{tc}_{p^\dagger}(Y))p(Y')$ (to prove $(\forall Y' \in \mathbf{tc}_{p^\dagger}(Z))p(Y')$). If $Y' \in \mathbf{tc}_{p^\dagger}(Z)$, then either $Y' \in Z$, so $p(Y')$ follows from $p^\dagger(Z)$, or $Y' \in \mathbf{tc}_{p^\dagger}(Y)$, where $Y \in Z$, so $p(Y')$ follows from the induction hypothesis. In part (ii) the induction hypothesis is $p^\dagger(Z) \wedge (\forall Y \in Z)(\mathbf{WF}(Y) \wedge \mathbf{tc}(Y) = \mathbf{tc}_{p^\dagger}(Y))$ to prove $\mathbf{WF}(Z) \wedge \mathbf{tc}(Z) = \mathbf{tc}_{p^\dagger}(Z)$. Clearly, $\mathbf{WF}(Z)$ follows from the induction hypothesis. Also, from the induction hypothesis we have $\mathbf{tc}(Z) = Z \cup \bigcup\{\mathbf{tc}(Y) : Y \in Z :\} = Z \cup \bigcup\{\mathbf{tc}_{p^\dagger}(Y) : Y \in Z\} = \mathbf{tc}_{p^\dagger}(Z)$. Part (iii) is immediate from (i) and (ii). □

Corollary 4.8.1 *Let p be a general set predicate and Z an arbitrary set. The following conditions are equivalent:*

(i) $p^\dagger(Z)$.

(ii) $\mathbf{WF}(Z) \wedge p(Z) \wedge (\forall Y \in \mathbf{tc}(Z))p(Y)$.

PROOF. The implication from (i) to (ii) is clear from Theorem 4.8. The implication from (ii) to (i) is proved by **WF**-induction on Z over the predicate $q(Z) \equiv p(Z) \wedge (\forall Y \in \mathbf{tc}(Z))p(Y) \rightarrow p^\dagger(Z)$. The induction hypothesis is $\mathbf{WF}(Z) \wedge (\forall Y \in Z)q(Y)$. To prove $q(Z)$ we assume $p(Z) \wedge (\forall Y \in \mathbf{tc}(Z))p(Y)$ (to prove $p^\dagger(Z)$). Now $p(Z)$ follows from the assumption. If $Y \in Z$ we have $q(Y)$ by the induction hypothesis and $p(Y)$ from the assumption on Z. Furthermore, if $Y' \in \mathbf{tc}(Y)$, then $\mathbf{tc}(Y') \subseteq \mathbf{tc}(Y)$ by Theorem 3.2, hence $p(Y')$. From $q(Y)$ we have $p^\dagger(Y)$, hence $p^\dagger(Z)$ by definition 4.1.7. □

EXERCISE 4.12 Let Z be a set. Prove the following conditions are equivalent:

(i) $\mathbf{HF}(Z)$.

(ii) $\mathbf{WF}(Z) \wedge \mathbf{FI}(\mathbf{tc}(Z))$.

4.2 Natural Numbers

We use set induction to introduce the natural number predicate. Although by this construction we get only a predicate, and not the set of natural numbers, this is sufficient to define the recursive numerical operations via the recursion rule in Chapter 5. The introduction of the set ω requires a different type of construction that is discussed in Chapter 7.

To represent the natural numbers we must determine a set that represents 0 (usually $\emptyset$) and an operation that generates the successor of a natural number. The usual approach is to generate the successor via the operation S. Since $S(Z)$ is a transitive set whenever Z is a transitive set, it follows that every natural number is a transitive set. Furthermore, $\bigcup S(Z) = Z$ whenever Z is a transitive set, so $\bigcup$ is the predecessor operation.

Informally, the natural numbers are the objects generated by the following process: $0 = \emptyset$ is a natural number; $1 = S(0) = \{0\}$ is a natural number;

$2 = S(1) = \{0, 1\}$ is a natural number; $3 = S(2) = \{0, 1, 2\}$ is a natural number; in general, if $n = Z$ is a natural number, then $n+1 = S(Z) = \{0, 1, ..., n\}$ is a natural number. The next definition gives an inductive characterization that captures the structure of these objects.

4.2.1 $\mathbf{NT}(Z) \equiv Z = \emptyset \vee (Z = S(\bigcup Z) \wedge \mathbf{NT}(\bigcup Z))$ (*Z is a natural number*).

4.2.2 $\mathbf{is}_{\mathbf{NT}}(Z) = \{\bigcup Z : Z = S(\bigcup Z) :\}$.

4.2.3 $\mathbf{tc}_{\mathbf{NT}}(Z) = [Z = \emptyset \Rightarrow \emptyset, \{\bigcup Z\} \cup \mathbf{tc}_{\mathbf{NT}}(\bigcup Z)]$.

4.2.4 $\mathbf{SC}(Z) \equiv (\exists Y \in Z) Z = S(Y)$ (*Z is a successor*).

4.2.5 $\mathbf{pd}(Z) = \bigcup Z$ (*the predecessor of Z*).

Note that 4.2.1 is exactly the definition of ind in Example 3.5. Clearly, $\emptyset$ is a natural number. Furthermore, $S(\emptyset) = S(\bigcup S(\emptyset))$ where $\bigcup S(\emptyset) = \emptyset$ is a natural number, so $S(\emptyset)$ is also a natural number. From the induction axiom it follows that if Z is a natural number, then either $Z = \emptyset$ or $Z = S(\bigcup Z)$. From now on we use the standard notation for natural numbers: $0 = \emptyset$, $1 = S(0) = \{0\}$, $2 = S(1) = \{0, 1\}$, $3 = S(2) = \{0, 1, 2\}$, etc.

EXAMPLE 4.1 Consider the predicate $p(Z) \equiv \mathbf{tc}_{\mathbf{NT}}(Z) = Z$. We show by **NT**-induction that $\mathbf{NT}(Z) \to p(Z)$. The induction hypothesis is $\mathbf{NT}(Z) \wedge (Z = 0 \vee p(\bigcup Z))$. The case $Z = 0$ is trivial, so we assume $Z = S(\bigcup Z)$, and by the induction hypothesis we have $\mathbf{tc}_{\mathbf{NT}}(\bigcup Z) = \bigcup Z$. We conclude that $\mathbf{tc}_{\mathbf{NT}}(Z) = \{\bigcup Z)\} \cup \bigcup Z = S(\bigcup Z) = Z$. □

EXERCISE 4.13 Let V be a set such that there is a natural number $Z \in V$. Prove there is a natural number $Z' \in V$ such that $V \cap Z' = \emptyset$.

EXERCISE 4.14 Let Z be a natural number. Prove that $Z \notin Z$.

Remark 4.4 The inductive tree induced by Definition 4.2.1 takes a particularly simple form. The tree for a given set Z consists of only one branch, where a node Z' has an immediate successor $\bigcup Z'$ if and only if $Z' = S(\bigcup Z')$. It follows that $\mathbf{NT}(Z)$ holds if and only if the branch halts at the node $\emptyset$. By reversing the branch we get the sequence $0, 1, ..., Z$. This means that every natural number is generated by the process described above. □

Remark 4.5 The general form of the induction hypothesis in a proof by **NT**-induction is $\mathbf{NT}(Z) \wedge (Z = \emptyset \vee (\forall Y \in \mathbf{tc}_{\mathbf{NT}}(Z))p(Y))$. On the other hand, from Example 4.1 it follows that whenever Z is a natural number, then $Z = \emptyset \vee (\forall Y \in \mathbf{tc}_{\mathbf{NT}}(Z))p(Y) \equiv (\forall Y \in Z)p(Y)$. Now from this equivalence it is clear that the induction hypothesis is equivalent to:

$$\mathbf{NT}(Z) \wedge (\forall Y \in Z)p(Y).$$

This formulation of numerical induction is usually called *complete induction.* In fact, it is a very convenient rule that can be used in many situations. Still, we prefer another form of induction that is somewhat weaker but is very natural. This is the rule of *mathematical induction*, which we proceed to define formally. □

Rule of Mathematical Induction Let p be a unary general predicate such that the following two conditions are satisfied by any set Z:

MI 1 $p(0)$.

MI 2 $(\mathbf{NT}(Z) \wedge p(Z)) \rightarrow p(S(Z))$.

Then $\mathbf{NT}(Z) \rightarrow p(Z)$ holds for any arbitrary set Z.

The validity of this rule can be proved by using **NT**-induction, but we must change the predicate p to a predicate $p'(Z) \equiv (\forall Y \in S(Z))p(Y)$ and prove by **NT**-induction that $\mathbf{NT}(Z) \rightarrow p'(Z)$ (note that $p'(Z)$ always implies $p(Z)$). The induction hypothesis is $(\forall Y \in Z)p'(Y)$. To prove $p'(Z)$, note that if $Z = 0$, then $p'(Z)$ follows from **MI 1**. Otherwise we have $p(Y)$ whenever $Y \in Z$ by the induction hypothesis and $Z = S(\bigcup Z)$ so $p(\bigcup Z)$. By **MI 2** it follows $p(Z)$, so $p'(Z)$.

Theorem 4.9 *Assume Z is a natural number. ,Then:*

(i) *Z is transitive.*

(ii) $\mathbf{pd}(S(Z)) = Z$.

(iii) *$S(Z)$ is a natural number.*

(iv) $Z = 0 \vee (\mathbf{SC}(Z) \wedge 0 \in Z)$.

(v) *Z is hereditarily finite.*

(vi) *If $Y \in Z$, then $Y \subset Z$ and Y is a natural number.*

PROOF. Part (i) is proved by mathematical induction. If $Z = \emptyset$, Z is clearly transitive. If Z is transitive,then $S(Z)$ is transitive. Part (ii) follows from the definitions, noting that by part (i) Z is transitive. To prove (iii) we use mathematical induction. Clearly, $S(\emptyset)$ is a natural number. If $S(Z)$ is a natural number, then $S(S(Z))$ is also a natural number by definition 4.2.1. Part (iv) follows by mathematical induction. In part (v) we prove that Z is hereditarily finite by mathematical induction. The case $Z = \emptyset$ is trivial. If Z is hereditarily finite, then clearly $S(Z)$ is hereditarily finite. Part (vi) follows by mathematical induction using the definitions. □

Corollary 4.9.1 *If Z and Z' are natural numbers and $S(Z) = S(Z')$, then $Z = Z'$.*

PROOF. We have $Z = \bigcup S(Z) = \bigcup S(Z') = Z'$. □

Corollary 4.9.2 *If Z be a set such that $(\exists Y \in Z)(Z = S(Y) \wedge \mathbf{NT}(Y))$, then Z is a natural number.*

PROOF. Since Y is a natural number we have $\bigcup Z = \bigcup S(Y) = Y$. It follows that $Z = S(\bigcup Z)$, where $\bigcup Z$ is a natural number. From definition 4.2.1 it follows that Z is a natural number. □

Corollary 4.9.3 *Let Z be a set. The following conditions are equivalent:*

(i) *Z is a natural number.*

(ii) $\mathbf{WF}(Z) \wedge \mathbf{TR}(Z) \wedge (Z = \emptyset \vee (\mathbf{SC}(Z) \wedge (\forall Y \in Z)(\mathbf{TR}(Y) \wedge (Y = \emptyset \vee \mathbf{SC}(Y)))))$.

PROOF. The implication from (i) to (ii) follows from Theorem 4.9. To prove the implication from (ii) to (i) we use **WF**-induction on Z over the predicate $p(Z) \equiv q(Z) \rightarrow \mathbf{NT}(Z))$, where the predicate q is defined,

$$q(Z) \equiv \mathbf{TR}(Z) \wedge (Z = \emptyset \vee (\mathbf{SC}(Z) \wedge (\forall Y \in Z)(\mathbf{TR}(Y) \wedge (Y = \emptyset \vee \mathbf{SC}(Y))))).$$

The induction hypothesis is that Z is well-founded and $p(Y)$ holds whenever $Y \in Z$. To prove $p(Z)$ we assume $q(Z)$. The case $Z = 0$ is trivial. Otherwise, there is $Y \in Z$ and from the assumptions it follows $q(Y)$, so by the induction hypothesis we have $\mathbf{NT}(Y)$. From Corollary 4.9.2 we get $\mathbf{NT}(Z)$. □

Remark 4.6 The preceding corollary shows that we can define explicitly the predicate **NT** from the predicate **WF**. In this case the rule of **NT**-induction can be derived by **WF**-induction. This can be useful in systems where the only form of induction is provided by the predicate **WF**. In the system $\mathbb{G}$ induction is a general procedure derived from the induction rule, and this reduction is not necessary. □

Theorem 4.10 *Let X and Z be natural numbers. , then:*

(i) $X \in Z \rightarrow S(X) = Z \vee S(X) \in Z$.

(ii) $X \notin Z \rightarrow X = Z \vee Z \in X$.

(iii) $X \in Z \vee X = Z \vee Z \in X$.

(iv) $X \subseteq Z \vee Z \subseteq X$.

(v) $X \in Z \equiv X \subset Z$.

(vi) $X \in S(Z) \equiv X \subseteq Z$.

PROOF. The proof of (i) is by mathematical induction on Z with X fixed. The case $Z = 0$ is trivial. Assume the property is true for the natural number Z, and to prove $S(Z)$ assume $X \in S(Z)$. This induces two cases, $X \in Z$, hence by the induction hypothesis we get $S(X) \in S(Z)$ and $X = Z$, hence $S(X) = S(Z)$. The proof of (ii) is also by mathematical induction on Z with X fixed. If $Z = \emptyset$ we use Theorem 4.9 (iv). If $X \notin S(Z)$, then $X \notin Z \wedge X \neq Z$, so by the induction hypothesis we have $X = Z \vee Z \in X$. Using part (i) we get $X = S(Z) \vee S(Z) \in X$. Part (iii) is a reformulation of (ii). Part (iv) follows from (iii) and Theorem 4.9 (vi). To prove (v) from left to right we again use Theorem 4.9 (vi). From right to left it follows from part (iii). Part (vi) follows from (v). □

The results in Theorem 4.10 show that $\subseteq$ is the natural order relation $\leq$ between natural numbers, and $\in$ is the $<$ relation. Part (v) of the theorem relates the two relations in the usual way. We can formalize these relations in the following notation.

4.2.6 $X \leq_{\mathbf{nt}} Z \equiv X \subseteq Z$ (*X is less or equal than Z*).

4.2.7 $X <_{\mathbf{nt}} Z \equiv X \in Z$ (*X is less than Z*).

With this notation the properties in Theorem 4.10 can be expressed in the following form, where X and Z are natural numbers:

$$X <_{\mathbf{nt}} Z \rightarrow S(X) = Z \vee S(X) <_{\mathbf{nt}} Z.$$

$$X \not<_{\mathbf{nt}} Z \rightarrow X = Z \vee Z <_{\mathbf{nt}} X.$$

$$X <_{\mathbf{nt}} Z \vee Z = Z \vee Z <_{\mathbf{nt}} X.$$

$$X \leq_{\mathbf{nt}} Z \vee Z \leq_{\mathbf{nt}} X.$$

$$X <_{\mathbf{nt}} Z \equiv X \subset Z.$$

$$X <_{\mathbf{nt}} S(Z) \equiv X \leq_{\mathbf{nt}} Z.$$

EXERCISE 4.15 Prove that if X and Z are natural numbers and $X <_{\mathbf{nt}} Z$, then $S(X) <_{\mathbf{nt}} S(Z)$.

EXERCISE 4.16 Prove that if X and Y are natural numbers and $X \leq_{\mathbf{nt}} Z$, then $S(X) \leq_{\mathbf{nt}} S(Z)$.

We now consider sets of natural numbers, in particular finite sets. If X is a set of natural numbers, the operations $\bigcup X$ and $\bigcap X$ have been defined in Chapter 2.

Theorem 4.11 *Let X be a non-empty set of natural numbers. Then:*

(i) $\bigcap X \in X$ *and* $(\forall Y \in X) \bigcap X \subseteq Y$.

(ii) *If X is finite, then* $\bigcup X \in X$, $(\forall Y \in X) Y \subseteq \bigcup X$ *and* $S(\bigcup X) = \bigcup\{S(Y) : Y \in X :\}$

(iii) *If X is finite and transitive, then* $X = S(\bigcup X)$ *and X is a natural number.*

PROOF. In part (i) we note that if $\bigcap X \notin X$, then by Theorem 4.10 (v) we have $\bigcap X \in Y$ for every $Y \in X$. This means $\bigcap X \in \bigcap X$, and this is a contradiction. In part (ii) the set X has a maximal element Y' in the sense of Theorem 4.5, and by Theorem 4.10 (iv) it follows that $\bigcup X = Y'$. On the other hand, $S(Y) \subseteq S(\bigcup X)$ for every $Y \in X$, hence $S(\bigcup X) = \bigcup\{S(Y) : Y \in X :\}$. In part (iii) we have $X \subseteq S(\bigcup X)$ from the definitions. To prove the converse, assume $Y \in S(\bigcup X)$. If $Y \in \bigcup X$, then by transitivity of X we have $Y \in X$. If $Y = \bigcup X$, then $Y \in X$ by part (ii). □

Remark 4.7 Note that when X is a finite set, then the results of Theorem 4.11 (i) and (ii) can be expressed in the form:

$$\bigcap X \in X \wedge (\forall Y \in X) \bigcap X \leq_{\mathbf{nt}} Y$$
$$\bigcup X \in X \wedge (\forall Y \in X) Y \leq_{\mathbf{nt}} \bigcup X$$

using definitions 4.2.6 and 4.2.7. □

EXERCISE 4.17 Let X be a set. Prove that the following conditions are equivalent:

(i) X is a transitive hereditarily finite set, where all the elements are transitive.

(ii) X is a natural number.

Remark 4.8 There are alternative ways to define inductively the predicate **NT**. For example, we can set

$$\mathbf{NT}'(Z) \equiv Z = \emptyset \lor ((\exists Y \in Z)Z = S(Y) \land (\forall Y \in Z)\mathbf{NT}'(Y)).$$

It can be shown that **NT** and **NT′** are equivalent. Still, the definition of **NT′** is a bit redundant, and we prefer the expression in 4.2.1. In fact, the **NT**-tree and the **NT′**-tree for a natural number Z are very different, although both contain the same information, namely, the elements of the set Z. □

EXERCISE 4.18 Let **NT′** be the predicate in Remark 4.7. Prove:

(a) If Z is an arbitrary set, then $\mathbf{NT}(Z) \equiv \mathbf{NT}'(Z)$.

(b) If Z is a natural number, then $\mathbf{is}_{\mathbf{NT}'}(Z) = Z$.

4.3 Well-Founded Relations

In Section 3.3 we characterized the well-founded elements relative to an ordering determined by a set operation g (or equivalently, extensionally induced by a set predicate p). It turned out that those well-founded elements are exactly the sets satisfying an inductive predicate $\mathbf{WF}_g$. In this section we consider the same problem in terms of relations rather than predicates, where a relation, as defined in Chapter 2, is a set of ordered pairs, usually denoted with a letter R. With this understanding we shall write XRY for $<X,Y> \in R$. Note that the results below are valid even if R is not a relation.

4.3.1 $\mathbf{fd}(R) = \mathbf{do}(R) \cup \mathbf{ra}(R)$ (*the field of R*).

4.3.2 $\mathbf{pp}(X, R) = \{Y \in \mathbf{fd}(R) :: YRX \wedge Y \neq X\}$ (*the set of proper predecessors of X in R*).

4.3.3 $\mathbf{WF}^*(X, R) \equiv (\forall Y \in \mathbf{pp}(X, R))\mathbf{WF}^*(Y, R)$ (*X is R-well-founded*).

Note that $\mathbf{WF}^*$ is an extensional induction; in fact $\mathbf{WF}^* = \mathbf{WF}^*_{\mathbf{pp}}$, and $\mathbf{is}_{\mathbf{WF}^*}(X, R) = \mathbf{pp}(X, R)$. We set $\mathbf{tc}^*(X, R) = \mathbf{tc}^*_{\mathbf{pp}}(X, R)$, so $\mathbf{pp}(X, R) \subseteq \mathbf{tc}^*(X, R)$. Note that X is R-well founded if and only if X is **pp**-well-founded at R. This means that the general discussion in Section 3.3 applies to the predicate $\mathbf{WF}^*$.

Remark 4.9 The representation of relations by means of sets containing ordered pairs is a very useful device, although it is artificial and in some situations generates ambiguities. For example, the introduction of *reflexive pairs* of the form $<X, X>$ is sometimes an inconvenience, and in other cases a necessity. Definitions 4.3.2 and 4.3.3 indicate that what really matters in a relation R are the non-reflexive pairs, of the form $<X, Y>$ with $X \neq Y$. Still, a relation $R = \{<X, X>\}$ is a proper relation where $\mathbf{fd}(R) = \{X\}$ and X is R-well-founded. If we eliminate the reflexive pair $<X, X>$ we get $R' = \emptyset$ which is a very different relation, in the sense that $\mathbf{fd}(R') = \emptyset$ (note that X is also R'-well-founded). We have to compromise, and in some cases we enforce reflexivity by requiring $<X, X> \in R$ for every $X \in \mathbf{fd}(R)$ (see definition 4.3.4). If we want to ignore the reflexive elements we introduce the following notation:

$$R' = \{<X, Y> : X\mathbf{fd}(R) \wedge Y \in \mathbf{fd}(R) : X \in \mathbf{pp}(Y, R)\}.$$

Again, note that if $R = \{<X, X>\}$, then $R' = \emptyset$. On the other hand, if $R = \{<X, X>, <X, Y>\}$ where $X \neq Y$, then $R' = \{<X, Y>\}$. On the other hand we have $\mathbf{fd}(R) = \mathbf{fd}(R') = \{X, Y\}$. A relation R is *trivial* if $R' = \emptyset$. In particular $R = \emptyset$ is trivial. □

EXERCISE 4.19 Prove that if $X \notin \mathbf{fd}(R)$, then X is R-well-founded and $\mathbf{pp}(X,R) = \mathbf{tc}^*(X,R) = \emptyset$.

EXERCISE 4.20 Prove that $\mathbf{tc}^*(X,R) \subseteq \mathbf{fd}(R)$.

EXERCISE 4.21 Prove that if $Y \in \mathbf{tc}^*(X,R)$, then $\mathbf{tc}^*(Y,R) \subset \mathbf{tc}^*(X,R)$.

EXERCISE 4.22 Prove that if $R \subseteq R'$, then $\mathbf{fd}(R) \subseteq \mathbf{fd}(R')$, $\mathbf{pp}(X,R) \subseteq \mathbf{pp}(X,R')$, and $\mathbf{tc}^*(X,R) \subseteq \mathbf{tc}^*(X,R')$.

EXERCISE 4.23 Prove that if $R' \subseteq R$ and X is R-well-founded, then X is R'-well-founded.

Theorem 4.12 *Assume the set X is R-well-founded. Then:*

(i) $X \notin \mathbf{tc}^*(X,R)$.

(ii) *If* $Y \in \mathbf{tc}^*(X,R)$, *then* $X \notin \mathbf{tc}^*(Y,R)$.

(iii) *If* $Y \in \mathbf{tc}^*(X,R)$ *and* $Z \in \mathbf{tc}^*(Y,R)$, *then* $X \notin \mathbf{tc}^*(Z,R)$.

PROOF. Immediate from Theorem 3.5. □

4.3.4 $\mathbf{sg}(X,R) = \{<Z,Y> : Z \in \mathbf{tc}^*(X,R) \wedge Y \in \mathbf{tc}^*(X,R) : Z = Y \vee ZRY\}$ *(the segment of R induced by X).*

If $\mathbf{tc}^*(X,R) = \emptyset$, then $\mathbf{sg}(X,R) = \emptyset$. If $\mathbf{tc}^*(X,R) \neq \emptyset$ we know that there is $X' \in \mathbf{tc}^*(X,R)$ such that $\mathbf{tc}^*(X',R) = \emptyset$, hence $\mathbf{sg}(X',R) = \emptyset$. If $\mathbf{tc}^*(X,R) = \{X'\}$, then $\mathbf{sg}(X,R) = \{<X',X'>\}$ (see Remark 4.9).

Theorem 4.13 *Assume X is R-well-founded. , then:*

(i) $\mathbf{fd}(\mathbf{sg}(X,R)) = \mathbf{tc}^*(X,R)$.

(ii) *If* $Y \in \mathbf{tc}^*(X,R)$, *then* $\mathbf{pp}(Y,R) = \mathbf{pp}(Y,\mathbf{sg}(X,R))$.

(iii) *If* $Y \in \mathbf{tc}^*(X,R)$, *then* $\mathbf{tc}^*(Y,R) = \mathbf{tc}^*(Y,\mathbf{sg}(X,R))$.

(iv) *If* $Y \in \mathbf{tc}^*(X,R)$, *then* $\mathbf{sg}(Y,R) = \mathbf{sg}(Y, \mathbf{sg}(X,R))$.

PROOF. Clearly, any element of $\mathbf{fd}(\mathbf{sg}(X,R))$ is an element of $\mathbf{tc}^*(X,R)$. On the other hand, if $Y \in \mathbf{tc}^*(X,R)$, then $<Y,Y> \in \mathbf{sg}(X,R)$, hence $Y \in \mathbf{fd}(\mathbf{sg}(X,R))$. Part (ii) follows immediately from the definitions. To prove (iii) we use $\mathbf{WF}^*$-induction on Y, over the predicate $p(Y,R) \equiv Y \in \mathbf{tc}^*(X,R) \to \mathbf{tc}^*(Y,R) = \mathbf{tc}^*(Y, \mathbf{sg}(X,R))$, with X fixed. The induction hypothesis is that $p(Y',R)$ holds whenever $Y' \in \mathbf{pp}(Y,R)$. To prove $p(Y,R)$, we assume $Y \in \mathbf{tc}^*(X,R)$ to prove $\mathbf{tc}^*(X,R) = \mathbf{tc}^*(Y, \mathbf{sg}(X,R))$. From this assumption it follows that whenever $Y' \in \mathbf{pp}(Y,R)$, then $Y' \in \mathbf{tc}^*(X,R)$, so by the induction hypothesis we have $\mathbf{tc}^*(Y',R) = \mathbf{tc}^*(Y', \mathbf{sg}(X,R))$. Now the conclusion follows using the closure axiom in the induction rule, the induction hypothesis, and part (ii). Finally, part (iv) follows using Definition 4.3.4 and part (iii). □

4.4 Notes

The inductive definition of finite sets comes from Whitehead and Russell (1912), where it is expressed using a highly impredicative language. This approach is still followed by contemporary writers (see Levy (1979)).

As shown in the text, the definition of hereditarily finite sets can be generalized to a predicate p. This alternative is applied later in several places, particularly the definition of the ordinals in Chapter 6 and in the definition of the hereditarily α-countable sets in the Appendix.

The natural numbers are usually defined as the elements of the set ω, a set that is introduced by a special axiom of infinity. Here the set ω is not available, although it is introduced later in Chapter 7. Still, the predicate provides a very useful machinery that is independent of the assertion of the set ω.

The theory of R-well-founded sets identifies those sets in the field of a relation R that are well-founded in the sense explained in Section 3.3. This definition does not imply that every element of the field is R-well-founded. Those elements that are well-founded induce a partial sub-relation of R where induction and recursion are available. In fact, the elements of the field that are not well-founded can be ignored. For example, the definition of the segment induced by a well-founded element X is independent of the non-well-founded elements. Relations that satisfy some extra condition of connectedness are studied in Section 5.2.

Chapter 5

Set Recursion

In this chapter the set induction rule is extended with a new rule that allows for the use of recursion to define new set operations. Each application of the recursion rule must be supported by some inductive predicate that determines the domain of the operation. As a consequence the rule of proof by induction can be used to prove properties of recursive operations.

5.1 Recursion

The rule of set recursion is used to define set operations on some underlying inductive predicate ind. Besides the predicate ind, the rule assumes that another set operation is given. As usual, the rule can be applied with basic or general set operations.

The Set Recursion Rule Let ind be a $(k+1)$-ary predicate defined by set induction from predicates q_1, q_2 and set operation g; and let h be a $(k+2)$-ary general set operation. We introduce a new $(k+1)$-ary set operation rec that satisfies the following axiom.

Set Recursion Axiom:

$$\mathsf{rec}(Z, \mathbf{X}) = [\mathsf{ind}(Z, \mathbf{X}) \Rrightarrow h((\lambda Y \in \mathbf{tc}_{\mathsf{ind}}(Z, \mathbf{X}))\mathsf{rec}(Y, \mathbf{X}), Z, \mathbf{X}), \emptyset].$$

Whenever the recursion rule is applied to introduce a set operation rec, and the underlying inductive predicate is ind, we say that rec is *defined by* ind-*recursion*, or simply *defined by recursion.*

The term $\emptyset$ provides a *global exit value* for the case when the condition $\mathsf{ind}(Z)$ fails. In applications we omit both the exit value and the condition $\mathsf{ind}(Z, \mathbf{X})$. With this understanding, the above axiom can be written:

$$\mathsf{rec}(Z, \mathbf{X}) = h((\lambda Y \in \mathbf{tc}_{\mathsf{ind}}(Z, \mathbf{X}))\mathsf{rec}(Y, \mathbf{X}), Z, \mathbf{X}).$$

Remark 5.1 If ind is a basic predicate and h is a basic operation, then by definition rec is a basic operation. If this is not the case we proceed as in Remark 3.1 and determine the list $\mathbf{V} = V_1, ..., V_s$ of all fixed sets that are required to write q_1, q_2, g, h as substitutions in basic q_1', q_2', g', h', and we introduce a new inductive predicate ind′ as in Remark 3.1. Now we define by ind′-recursion the basic operation rec′ such that

$$\mathsf{rec}'(Z, \mathbf{X}, \mathbf{Y}) = h'((\lambda Y \in \mathbf{tc}_{\mathsf{ind}'}(Z, \mathbf{X}, \mathbf{Y}))\mathsf{rec}'(Y, \mathbf{X}, \mathbf{Y}), Z, \mathbf{X}, \mathbf{Y}),$$

where $\mathbf{Y}$ stands for new variables $Y_1, ..., Y_s$. It follows that $\mathsf{rec}(Z, \mathbf{X}) = \mathsf{rec}'(Z, \mathbf{X}, \mathbf{V})$, so rec is a general operation. □

EXERCISE 5.1 Prove, using ind-induction, that $\mathsf{rec}(Z, \mathbf{X}) = \mathsf{rec}'(Z, \mathbf{X}, \mathbf{V})$ in Remark 5.1 holds for arbitrary sets $Z, \mathbf{X}$.

EXAMPLE 5.1 In Chapter 3 we discussed a number of possible variants in the inductive definition of the predicate ind. Clearly, the set recursion rule is independent of such alternatives. In fact the operation rec is determined completely by ind and h. In applications, the form of the inductive definition is relevant. For example, consider a unary predicate ind of the form referred to in Remark 3.3, where the term W is omitted. This predicate is determined by the predicates q_1, q_2 and a term V via the axiom,

$$\mathsf{ind}(Z, \mathbf{X}) \equiv q_1(Z, \mathbf{X}) \vee (q_2(Z, \mathbf{X}) \wedge \mathsf{ind}(\mathrm{V}, \mathbf{X})).$$

We know this is equivalent to a standard induction of the form

$$\mathsf{ind}(Z,\mathbf{X}) \equiv q_1(Z,\mathbf{X}) \vee (q_2(Z,\mathbf{X}) \wedge (\forall Y \in \{\mathrm{V}\})\mathsf{ind}(Y,\mathbf{X})).$$

Hence, if h is a $(k+2)$-ary set operation we can introduce a $(k+1)$-ary set operation rec such that

$$\mathsf{rec}(Z,\mathbf{X}) = h((\lambda Y \in \mathbf{tc}_{\mathsf{ind}}(Z,\mathbf{X}))\mathsf{rec}(Y,\mathbf{X}), Z, \mathbf{X}).$$

Note that in this case $\mathbf{tc}_{\mathsf{ind}}(Z,\mathbf{X}) = \{\mathrm{V}\} \cup \mathbf{tc}_{\mathsf{ind}}(\mathrm{V},\mathbf{X})$. □

In applications we can replace the operation h by a set term V and write the recursive equation in the form

$$\mathsf{rec}(Z,\mathbf{X}) = \mathrm{V}.$$

We require, of course, that the free variables in the term V are in the list $Z, \mathbf{X}$. Furthermore, we allow in V free occurrences of the term $\mathrm{V}' = (\lambda Y \in \mathbf{tc}_{\mathsf{ind}}(Z,\mathbf{X}))\mathsf{rec}(Y,\mathbf{X})$. As usual, an occurrence is free when no binding operation in V affects free variables in V′. Also we allow in V free occurrences of expressions obtained by decoding V′. In this connection there are three operations, which have been defined in Chapter 2, that provide the basis of the decoding, namely, the operations **do**, **ra**, and **ap**. Their role comes from three identities, where again we ignore the parameters in the recursive operation. In fact, the term V′ is a function, and the following relations are valid:

$\mathbf{do}(\mathrm{V}') = \mathbf{tc}_{\mathsf{ind}}(Z)$.

$\mathbf{ra}(\mathrm{V}') = \{\mathsf{rec}(Y) : Y \in \mathbf{tc}_{\mathsf{ind}}(Z) :\}$.

$\mathbf{ap}(\mathrm{V}', Y) = \mathsf{rec}(Y)$ if $Y \in \mathbf{tc}_{\mathsf{ind}}(Z)$.

This means that all these expressions may occur in the term V. Still, it is difficult to give a general characterization of the structure of the term

V, and a great deal of caution is required to make sure that the definition can be brought to the proper form. The final test is the actual definition of the operations h in such a way that the crucial equation is satisfied: $h((\lambda Y \in \mathbf{tc}_{\mathsf{ind}}(Z,\mathbf{X}))\mathsf{rec}(Y,\mathbf{X}), Z, \mathbf{X}) = \mathrm{V}$. In many cases it is obvious how h is defined, and in others this may require extra information about the underlying inductive predicate ind.

EXAMPLE 5.2 The following example will clarify the preceding remark. We introduce a binary inductive predicate ind,

$$\mathsf{ind}(Z,X) \equiv Z = \emptyset \vee (Z = \{\bigcup Z\} \wedge (\forall Y \in Z)\mathsf{ind}(Y,X)).$$

We also define by ind-recursion the operation,

$$\mathsf{rec}(Z,X) = [Z = \emptyset \Rrightarrow X, \{\mathsf{rec}(\bigcup Z, X)\}].$$

Just by inspection it is not clear that this is a legitimate recursion, so we have to make sure that the 3-ary operation h can be defined in such a way that $h(\lambda Y \in \mathbf{tc}_{\mathsf{ind}}(Z,X))\mathsf{rec}(Y,X), Z, X) = [Z = \emptyset \Rrightarrow X, \{\mathsf{rec}(\bigcup Z, X)\}]$. We can show that this h can be defined, noting that under the assumption $\mathsf{ind}(Z,X) \wedge Z \neq \emptyset$, it follows that $\bigcup Z \in Z \subseteq \mathbf{tc}_{\mathsf{ind}}(Z)$, hence $\mathbf{ap}((\lambda Y \in \mathbf{tc}_{\mathsf{ind}}(Z,X))\mathsf{rec}(Y,X), \bigcup Z) = \mathsf{rec}(\bigcup Z, X)$, and we can take the set operation h in the form,

$$h(V,Z,X) = [Z = \emptyset \Rrightarrow X, \{\mathbf{ap}(V, \bigcup Z)\}].$$

The parameter X in the definition of ind is not essential in the sense of Remark 3.3. Note that the parameter is certainly essential for the definition of rec. □

EXERCISE 5.2 Let Z and X be arbitrary sets. Prove that $\mathsf{ind}'(Z) \equiv \mathsf{ind}(Z,X)$, where ind′ is the predicate in Example 3.6 and ind is the predicate in Example 5.2.

We proceed now to explain the primitive construction that supports the recursion rule. This construction is derived from the discussion of the induction rule in Section 3.2. Here we consider rec to be a unary operation, and the parameters **X** are omitted. This is done for convenience and does not affect the generality of the argument. We assume the recursion is of the form,

$$\mathsf{rec}(Z) = h((\lambda Y \in \mathbf{tc}_{\mathsf{ind}}(Z))\mathsf{rec}(Y), Z).$$

Recall that $\mathsf{ind}(Z)$ holds for some set Z if and only if every branch in the inductive tree is closed (or simply, if the inductive tree is closed).

If $\mathsf{ind}(Z)$ fails we set $\mathsf{rec}(Z) = \emptyset$. Otherwise, the inductive tree for Z is closed and we proceed to explain the evaluation of $\mathsf{rec}(Z')$ for every node Z' in the inductive tree for Z. If Z' is a closing node, this means that $\mathbf{tc}_{\mathsf{ind}}(Z') = \emptyset$ and we set $\mathsf{rec}(Z') = h(\emptyset, Z')$. Clearly, this is consistent with the recursive equation.

We consider now the case where the node Z' is not halting, and we assume that for every node Z'' in the sub-tree induced by Z', $\mathsf{rec}(Z'')$ has been evaluated in a form consistent with the recursive definition. This means that $\mathrm{V}' = (\lambda Y \in \mathbf{tc}_{\mathsf{ind}}(Z'))\mathsf{rec}(Y)$ is available, so we set $\mathsf{rec}(Z') = h(\mathrm{V}', Z')$, and this is consistent with the recursive equation.

As long as this process is continued we evaluate $\mathsf{rec}(Z')$ for nodes in the tree in a manner consistent with the recursive equation. We claim that every node in the tree is evaluated. In fact, if this is not the case and $\mathsf{rec}(Z')$ is not evaluated, then there is an immediate successor Z'' of Z' where $\mathsf{rec}(Z'')$ is not evaluated. This means that there is a non-halting branch in the tree, contradicting the definition of ind.

5.2 Rank

The first application of the recursion rule is a construction that associates with every $(k+1)$-ary inductive predicate ind a $(k+1)$-ary set operation

$\mathbf{rk}_{\mathsf{ind}}$ such that in case $\mathsf{ind}(Z, \mathbf{X})$ holds, then $\mathbf{rk}_{\mathsf{ind}}(Z, \mathbf{X})$ determines the complexity of the inductive tree for Z at $\mathbf{X}$. For example, if the inductive predicate is **FI** we want $\mathbf{rk}_{\mathbf{FI}}(Z, \mathbf{X})$ to be the number of elements in the set Z. Later we shall show that $\mathbf{rk}_{\mathsf{ind}}(Z, \mathbf{X})$ is always an ordinal, in the sense to be explained in Chapter 6.

The operation $\mathbf{rk}_{\mathsf{ind}}$ is defined by ind-recursion.

5.2.1 $\mathbf{rk}_{\mathsf{ind}}(Z, \mathbf{X}) = \{\mathbf{rk}_{\mathsf{ind}}(Y, \mathbf{X}) : Y \in \mathbf{tc}_{\mathsf{ind}}(Z, \mathbf{X}) :\}$.

This means that the operation h required by the recursive equation is defined $h(V, Z, \mathbf{X}) = \mathbf{ra}(V)$ (see also the discussion in Example 5.1). Note that if $\mathsf{ind}(Z, \mathbf{X})$ fails, then $\mathbf{rk}_{\mathsf{ind}}(Z, \mathbf{X}) = \emptyset$ by default.

Theorem 5.1 *Let $Z, \mathbf{X}$ be arbitrary sets. Then:*

(i) $\mathbf{rk}_{\mathsf{ind}}(Z, \mathbf{X})$ *is well-founded.*

(ii) $(\forall Y \in \mathbf{tc}_{\mathsf{ind}}(Z, \mathbf{X}))\mathbf{rk}_{\mathsf{ind}}(Y, \mathbf{X}) \in \mathbf{rk}_{\mathsf{ind}}(Z, \mathbf{X})$.

(iii) $(\forall Y \in \mathbf{tc}_{\mathsf{ind}}(Z, \mathbf{X}))\mathbf{rk}_{\mathsf{ind}}(Y, \mathbf{X}) \subset \mathbf{rk}_{\mathsf{ind}}(Z, \mathbf{X})$.

(iv) $\mathbf{rk}_{\mathsf{ind}}(Z, \mathbf{X})$ *is transitive.*

PROOF. The theorem is trivial in case $\neg\mathsf{ind}(Z, \mathbf{X})$. Part (i) is proved by ind-induction. From the induction hypothesis it follows that the elements of $\mathbf{rk}_{\mathsf{ind}}(Z, \mathbf{X})$ are well-founded, so $\mathbf{rk}_{\mathsf{ind}}(Z, \mathbf{X})$ is also well-founded. Part (ii) is immediate from Definition 5.2.1. In part (iii) we note that if $Y \in \mathbf{tc}_{\mathsf{ind}}(Z, \mathbf{X})$, then by Corollary 3.2.1 we have $\mathbf{tc}_{\mathsf{ind}}(Y, \mathbf{X}) \subset \mathbf{tc}_{\mathsf{ind}}(Z, \mathbf{X})$, hence from Definition 5.2.1 we have $\mathbf{rk}_{\mathsf{ind}}(Y, \mathbf{X}) \subseteq \mathbf{rk}_{\mathsf{ind}}(Z, \mathbf{X})$. Furthermore, $\mathbf{rk}_{\mathsf{ind}}(Y, \mathbf{X}) \notin \mathbf{rk}_{\mathsf{ind}}(Y, \mathbf{X})$ by part (i), so we have $\mathbf{rk}_{\mathsf{ind}}(Y, \mathbf{X}) \subset \mathbf{rk}_{\mathsf{ind}}(Z, \mathbf{X})$. Part (iv) follows from (iii). □

EXERCISE 5.3 Let Z be a finite set. Prove:

(a) If $Z = \emptyset$, then $\mathbf{rk}_{\mathbf{FI}}(Z) = \emptyset$.

(b) If $Y \notin Z$, then $\mathbf{rk_{FI}}(Z;Y) = S(\mathbf{rk_{FI}}(Z))$.

(c) $\mathbf{rk_{FI}}(Z)$ is a natural number.

EXERCISE 5.4 Let Z be a natural number. Prove:

(a) $\mathbf{rk_{NT}}(Z) = Z$.

(b) $\mathbf{rk_{FI}}(Z) = Z$.

In dealing with well-founded sets we use a special notation. Recall that in Chapter 3 we set $\mathbf{tc}(Z) = \mathbf{tc_{WF}}(Z)$.

5.2.2 $\mathbf{rk}(Z) = \mathbf{rk_{WF}}(Z)$ (*the rank of Z*).

Note that $\mathbf{TR}(Z) \equiv \mathbf{TR_{WF}}(Z)$ holds for any well-founded set Z (Theorem 3.6 (iii)). Furthermore, if Z is well-founded, then $\mathbf{is_{WF}}(Z) = Z$.

We proceed now to study the operations of transitive closure and rank in the context of the predicate $\mathbf{WF}^*$ from Section 4.3.

5.2.3 $\mathbf{rk}^*(X,R) = \mathbf{rk_{WF^*}}(X,R)$ (*the relational rank of X in R*).

5.2.4 $\mathbf{rk}^{**}(R) = \{\mathbf{rk}^*(X,R) : X \in \mathbf{fd}(R) : \mathbf{WF}^*(X,R)\}$ (*the relational rank of R*).

Note that neither 5.2.3 nor 5.2.4 is a recursive definitions.

EXERCISE 5.5 Let Z be a well-founded set and define a relation R in the form,

$$R = \{<Y,X> : Y \in \mathbf{tc}(Z) \wedge X \in \mathbf{tc}(Z) : Y \in X \vee Y = X\}.$$

Prove:

(a) $\mathbf{fd}(R) = \mathbf{tc}(Z)$.

(b) $(\forall X \in \mathbf{fd}(R))\mathbf{WF}^*(X,R)$.

(c) $(\forall X \in \mathbf{fd}(R))\mathbf{rk}^*(X,R) = \mathbf{rk}(X)$.

(d) $\mathbf{rk}^{**}(R) = \mathbf{rk}(Z)$.

Theorem 5.2 *Assume R is an arbitrary relation. Then:*

(i) $\mathbf{rk}^{**}(R)$ *is well-founded.*

(ii) $\mathbf{rk}^{**}(R)$ *is transitive.*

PROOF. Part (i) is clear from Definition 5.2.4 and Theorem 5.1 (i). Part (ii) is clear from Definitions 5.2.1 and 5.2.3. □

5.2.5 $\mathbf{CO}(R) \equiv (\forall X \in \mathbf{fd}(R))(\forall Y \in \mathbf{fd}(R))(\mathbf{WF}^*(X,R) \wedge \mathbf{WF}^*(Y,R) \rightarrow (X = Y \vee X \in \mathbf{pp}(Y,R) \vee Y \in \mathbf{pp}(X,R))$ (*R is connected*).

5.2.6 $\mathbf{WO}(R) \equiv \mathbf{CO}(R) \wedge (\forall Y \in \mathbf{fd}(R))\mathbf{WF}(Y,R)$ (*R is well-ordered*).

EXERCISE 5.6 Assume R is connected and X, Y are both R-well-founded. Prove that if $X \neq Y$, then either XRY or YRX.

Theorem 5.3 *Assume R is connected, X is R-well-founded, and Y is a set. The following conditions are equivalent:*

(i) $Y \in \mathbf{pp}(X,R)$.

(ii) $Y \in \mathbf{tc}^*(X,R)$.

(iii) $\mathbf{rk}^*(Y,R) \in \mathbf{rk}^*(X,R) \wedge Y \in \mathbf{tc}^*(X,R)$.

PROOF. The implication from (i) to (ii) is trivial, since $\mathbf{pp}(X,R) \subseteq \mathbf{tc}^*(X, R)$. The implication from (ii) to (iii) follows from Definition 5.2.3. Assume (iii) and to prove (i) by contradiction, assume $Y \notin \mathbf{pp}(X,R)$. From (iii) and Theorem 5.1 (i) we know that $Y \neq X$. On the other hand, $X \in \mathbf{pp}(X,R)$ implies $\mathbf{rk}^*(X,R) \in \mathbf{rk}^*(Y,R)$ that again contradicts Theorem 5.1 (i). From Definition 5.2.5 we conclude that $Y \in \mathbf{pp}(X,R)$. □

EXERCISE 5.7 Assume R is connected and X is R-well-founded. Prove that $\mathbf{pp}(X,R) = \mathbf{tc}^*(X,R)$.

Remark 5.2 Assume R is a well-ordered relation. We introduce a function F such that

$$F = \{<Z, \mathbf{rk}^*(Z,R)> : Z \in \mathbf{fd}(R) :\}.$$

Clearly $\mathbf{do}(F) = \mathbf{fd}(R) \wedge \mathbf{ra}(F) = \mathbf{rk}^{**}(R)$, and $\mathbf{ap}(F,Z) = \mathbf{rk}^*(Z,R)$ whenever $Z \in \mathbf{fd}(R)$. Furthermore, the function F is 1–1. For, if $\mathbf{ap}(F,Y) = \mathbf{ap}(F,Z)$, then we have $Y = Z$. In fact, if this is not the case, since R is connected we have $Y \in \mathbf{pp}(Z,R)$ or $Z \in \mathbf{pp}(Y,R)$, hence by Theorem 5.3 we have $\mathbf{ap}(F,Y) \in \mathbf{ap}(F,Z)$ or $\mathbf{ap}(F,Z) \in \mathbf{ap}(F,Y)$. This is a contradiction, for $\mathbf{rk}^*(Z,R)$ is well-founded (Theorem 5.1). □

We have mentioned above that $\mathbf{rk}^*(Z,R)$ and $\mathbf{rk}^{**}(R)$ are ordinals. See applications of these results in Examples 6.4, 6.5, and 6.6.

5.3 Counting

We know that the **FI**-transitive closure of a finite set Z produces the set of all proper subsets of Z (Theorem 4.6). In this section we show that the **FI**-rank of Z is a natural number that counts the elements in Z. This being the case, it is convenient to introduce a special notation to make explicit this property of the rank operation.

5.3.1 $\mathbf{ct}(Z) = \mathbf{rk}_{\mathbf{FI}}(Z)$ (*the counting of Z*).

Note that we are setting implicitly $\mathbf{rk}_{\mathbf{FI}}(Z) = \emptyset$ when Z is not finite, although it would be more sensible to take some infinite set. This is not possible at the present stage of our theory.

EXAMPLE 5.3 Let $Z = \{X\}$. It follows that $\mathbf{tc}_{\mathbf{FI}}(Z) = \{\emptyset\}$, hence $\mathbf{ct}(Z) = \{0\} = 1$. If $Z = \{X, Y\}$, then $\mathbf{tc}_{\mathbf{FI}}(Z) = \{\emptyset, \{X\}, \{Y\}\}$, hence $\mathbf{ct}(Z) = \{0, 1\} = 2$. □

In the next theorem recall that whenever Y and Z are natural numbers, then $Y <_{\mathbf{nt}} Z$ means $Y \in Z$, and $Y \leq_{\mathbf{nt}} Z$ means $Y \subseteq Z$.

Theorem 5.4 *Let Z be a finite set, Y an arbitrary set, and V an arbitrary natural number. Then:*

(i) $\mathbf{ct}(Z)$ *is a natural number.*

(ii) *If $Y \subset Z$, then $\mathbf{ct}(Y) <_{\mathbf{nt}} \mathbf{ct}(Z)$.*

(iii) *If $V <_{\mathbf{nt}} \mathbf{ct}(Z)$, then $(\exists Y \in Z) V \leq_{\mathbf{nt}} \mathbf{ct}(Z \ominus Y)$.*

PROOF. Part (i) follows by **FI**-induction. By the induction hypothesis $\mathbf{ct}(Z)$ is a set of natural numbers, by Theorem 5.1 (iv) $\mathbf{ct}(Z)$ is transitive, and by Corollary 4.3.2 $\mathbf{ct}(Z)$ is a finite set. From Theorem 4.11 (iii) it follows that $\mathbf{ct}(Z)$ is a natural number. In part (ii) we note that if $Y \subset Z$, then by Theorem 4.6 $Y \in \mathbf{tc_{FI}}(Z)$, hence $\mathbf{ct}(Y) <_{\mathbf{nt}} \mathbf{ct}(Z)$. To prove part (iii) we note that $V = \mathbf{ct}(X)$ where $X \in \mathbf{tc_{FI}}(Z)$, and by Theorem 4.1 (iv) there is $Y \in Z$ such that $X \subseteq Z \ominus Y$. From part (ii) we conclude that $V \leq_{\mathbf{nt}} \mathbf{ct}(Z \ominus Y)$. □

Theorem 5.5 *Let V_0 be a fixed set. If Z is a finite set such that $V_0 \in Z$, then $(\forall V \in Z)\mathbf{ct}(Z \ominus V_0) \leq_{\mathbf{nt}} \mathbf{ct}(Z \ominus V)$.*

PROOF. The proof is by **FI**-induction over the predicate $p(Z) \equiv V_0 \in Z \rightarrow (\forall V \in Z)\mathbf{ct}(Z \ominus V_0) \leq_{\mathbf{nt}} \mathbf{ct}(Z \ominus V)$ with V_0 fixed. The relevant part of the induction hypothesis is that $p(Z \ominus V')$ holds whenever $V' \in Z$. To prove $p(Z)$ we assume $V_0 \in Z$, $V \in Z$ and $V \neq V_0$. We assume also $Y <_{\mathbf{nt}} \mathbf{ct}(Z \ominus V_0)$ (to prove $Y <_{\mathbf{nt}} \mathbf{ct}(Z \ominus V)$). From Theorem 5.4 (iii) we know there is $V' \in Z \ominus V_0$ such that $Y \leq_{\mathbf{nt}} \mathbf{ct}((Z \ominus V_0) \ominus V')$ and clearly $(Z \ominus V_0) \ominus V' = (Z \ominus V') \ominus V_0$. If $V = V'$, then by Theorem 5.4 (ii) we have $\mathbf{ct}((Z \ominus V) \ominus V_0) <_{\mathbf{nt}} \mathbf{ct}(Z \ominus V)$. If $V \neq V'$, then by the induction hypothesis we have $p(Z \ominus V')$, and since $V \in Z \ominus V'$ we have $Y <_{\mathbf{nt}} \mathbf{ct}((Z \ominus V') \ominus V) <_{\mathbf{nt}} \mathbf{ct}(Z \ominus V)$. □

EXERCISE 5.8 Let Z be a finite set, $V_0 \in Z$ and $V_1 \in Z$. Prove that $\mathbf{ct}(Z \ominus V_0) = \mathbf{ct}(Z \ominus V_1)$.

Corollary 5.5.1 *If Z is finite and $V \in Z$, then* $\mathbf{ct}(Z) = S(\mathbf{ct}(Z \ominus V))$.

PROOF. By Theorem 4.11 (iii) we know there is $Y \in \mathbf{tc_{FI}}(Z)$ such that $\mathbf{ct}(Z) = S(\mathbf{ct}(Y))$, and there is $V_0 \in Z$ such that $\mathbf{ct}(Y) \leq_{\mathbf{nt}} \mathbf{ct}(Z \ominus V_0)$. Hence, by Theorem 5.5 we have:

$$\mathbf{ct}(Z) = S(\mathbf{ct}(Y)) \leq_{\mathbf{nt}} S(\mathbf{ct}(Z \ominus V_0)) \leq_{\mathbf{nt}} S(\mathbf{ct}(Z \ominus V)) \leq_{\mathbf{nt}} \mathbf{ct}(Z).$$

This shows that $\mathbf{ct}(Z) = S(\mathbf{ct}(Z \ominus V))$, where $V \in Z$. □

Corollary 5.5.2 *If Z is finite and $V \notin Z$, then* $\mathbf{ct}(Z; V) = S(\mathbf{ct}(Z))$.

PROOF. Immediate from Corollary 5.5.1, noting that $(Z; V) \ominus V = Z$. □

EXAMPLE 5.4 Assume Z is a finite set. We would like to use **FI**-induction on Z to prove that there are functions F and F' such that $\mathbf{do}(F) = \mathbf{ra}(F') = Z$, $\mathbf{ra}(F) = \mathbf{do}(F') = \mathbf{ct}(Z)$, and whenever $Y \in Z$, then $\mathbf{ap}(F', \mathbf{ap}(F, Y))$. The problem is that the existential quantifiers on F and F' are non-local and cannot be used to express the induction predicate. To fix this situation we use Theorems 4.4 and 4.6, noting that the sets $Z \times \mathbf{ct}(Z)$ and $\mathbf{ct}(Z) \times Z$ are finite, so we can write $F \in \mathbf{tc_{FI}}(Z \times \mathbf{ct}(Z)) \cup \{Z \times \mathbf{ct}(Z)\}$ and $F' \in \mathbf{tc_{FI}}(\mathbf{ct}(Z) \times Z) \cup \{\mathbf{ct}(Z) \times Z\}$. Now we can proceed with the induction in the usual way. If $Z = \emptyset$ we take $F = F' = \emptyset$. Otherwise, there is $Y \in Z$. Let F_1 and F_1' be the functions satisfying the conditions for $Z \ominus Y$. We set $F = F_1 \cup \{< Y, \mathbf{ct}(Z \ominus Y) >\}$, and $F' = F_1' \cup \{< \mathbf{ct}(Z \ominus Y), Y >\}$, and clearly F, F' satisfy the required conditions. □

Theorem 5.6 *Let F be a 1–1 function and Z a natural number such that $Z \subseteq \mathbf{do}(F)$. Then $F[Z]$ is a finite set and* $\mathbf{ct}(F[Z]) = Z$.

PROOF. From Corollary 4.1.4 we know that $F[Z]$ is a finite set. We prove by mathematical induction on Z that whenever $Z \subseteq \mathbf{do}(F)$, then $\mathbf{ct}(F[Z]) = Z$.

The case $Z = \emptyset$ is trivial. Assume it is true for Z and $S(Z) \subseteq \mathbf{do}(F)$. It follows that $Z \subseteq S(Z) \subseteq \mathbf{do}(F)$, hence by the induction hypothesis we have $\mathbf{ct}(F[Z]) = Z$. Noting that $F[S(Z)] = F[Z]; \mathbf{ap}(F, Z)$, we have by Corollary 5.5.2 that $\mathbf{ct}(F[S(Z)]) = S(\mathbf{ct}(Z)) = S(Z)$. □

Corollary 5.6.1 *If F is a 1–1 function and $\mathbf{do}(F)$ is a natural number, then $\mathbf{ra}(F)$ is a finite set and $\mathbf{ct}(\mathbf{ra}(F)) = \mathbf{do}(F)$.*

PROOF. From Theorem 5.6, noting that $F[\mathbf{do}(F)] = \mathbf{ra}(F)$. □

5.4 Numerical Recursion

Applications of **NT**-recursion are called *numerical recursion*. A set operation introduced by numerical recursion is called a *numerical operation*. In principle, applications of numerical recursion follow the general pattern formulated in the set recursion rule as defined in Section 5.1. Such a recursion is determined by a binary set operation h, and satisfies the axiom,

$$\mathsf{rec}(Z) = h((\lambda Y \in Z)\mathsf{rec}(Y), Z),$$

where we take advantage of the relation $\mathbf{tc_{NT}}(Z) = Z$ when Z is a natural number (see Example 4.1).

It is convenient to reformulate the rule of **NT**-recursion in the form of a rule of primitive recursion that follows the pattern of the rule of mathematical induction.

Rule of Primitive Recursion Let r be a k-ary set operation, and h a $(k+2)$-ary set operation. We introduce a $(k+1)$-ary set operation rec that satisfies the following axioms where Z is a natural number:

PR 1 $\mathsf{rec}(0, \mathbf{X}) = r(\mathbf{X})$.

PR 2 $\mathsf{rec}(S(Z), \mathbf{X}) = h(\mathsf{rec}(Z, \mathbf{X}), Z, \mathbf{X})$.

Note that the rule requires two set operations, r and h. Furthermore, the recursive operation rec is $(k+1)$-ary. This means that we are implicitly extending the induction by adding k parameters. With the same notation **NT**, the induction axiom takes the form,

$$\mathbf{NT}(Z, \mathbf{X}) \equiv Z = 0 \vee \left(Z = S\left(\bigcup Z\right) \wedge \mathbf{NT}(Z, \mathbf{X})\right).$$

Now we introduce the operation rec by **NT**-recursion in the form,

$$\mathsf{rec}(Z, \mathbf{X}) = h'((\lambda Y \in Z)\mathsf{rec}(Y, \mathbf{X}), Z, \mathbf{X}),$$

where $h'(V, Z, \mathbf{X}) = [Z = \emptyset \Rightarrow r(\mathbf{X}), h(\mathbf{ap}(V, \bigcup Z), Z, \mathbf{X})]$. It is easy to verify that the set operation rec defined in this way satisfies the axioms **PR 1** and **PR 2**.

Most applications of **NT**-recursion will be in the form prescribed by the rule of primitive recursion. In practice, instead of operations r and h we may use terms. If $k = 0$, then $r(\mathbf{X})$ becomes a set from the universe.

Now we proceed to define by primitive recursion a binary operation **fs** that generates all the finite subsets of a given set W.

5.4.1-1 $\mathbf{fs}(0, W) = \{\emptyset\}$.

5.4.1-2 $\mathbf{fs}(S(Z), W) = \{X; Y : X \in \mathbf{fs}(Z, W) \wedge Y \in W :\} \cup \{\emptyset\}$.

EXAMPLE 5.5 Note that $\mathbf{fs}(1, W) = \{\{Y\} : Y \in W :\} \cup \{\emptyset\}$, where W is an arbitrary set. It follows that $\mathbf{fs}(1, \emptyset) = \{\emptyset\}$, and in general $\mathbf{fs}(Z, \emptyset) = \{\emptyset\}$ holds for every natural number Z. Also, $\mathbf{fs}(0, \{X, Y\}) = \{\emptyset\}$, $\mathbf{fs}(1, \{X, Y\}) = \{\emptyset, \{X\}, \{Y\}\}$ and $\mathbf{fs}(2, \{X, Y\}) = \{\emptyset, \{X\}, \{Y\}, \{X, Y\}\}$. In general, $\emptyset \in \mathbf{fs}(Z, W)$ for every natural number Z and set W. □

EXERCISE 5.9 Identify the set operation h such that **fs** can be defined by **NT**-recursion in the form $\mathbf{fs}(Z, W) = h((\lambda Y \in Z)\mathbf{fs}(Y, W), Z, W)$.

Theorem 5.7 *If Z is a natural number and W is a set, then* $\mathbf{fs}(Z, W) \subseteq \mathbf{fs}(S(Z), W)$.

PROOF. The proof is by mathematical induction on Z. The case $Z = 0$ is trivial. To prove $\mathbf{fs}(S(Z), W) \subseteq \mathbf{fs}(S(S(Z)), W)$ we assume $X \in \mathbf{fs}(S(Z), W)$ and $X \neq \emptyset$. This means $X = X'; Y$, where $X' \in \mathbf{fs}(Z, W)$ and $Y \in W$. By the induction hypothesis it follows that $X' \in \mathbf{fs}(S(Z), W)$, hence $X \in \mathbf{fs}(S(S(Z)), W)$. □

Corollary 5.7.1 *If Z, Z' are natural numbers and $Z' \leq_{\mathbf{nt}} Z$, then* $\mathbf{fs}(Z', W) \subseteq \mathbf{fs}(Z, W)$.

PROOF. By mathematical induction on Z with Z' fixed. The case $Z = 0$ is trivial. We assume that the assertion is true for Z, and also that $Z' <_{\mathbf{nt}} S(Z)$. Hence $Z' \leq_{\mathbf{nt}} Z$, and we use the induction hypothesis and Theorem 5.7. □

Theorem 5.8 *Let Z be a natural number and W a finite (hereditarily finite) set. Then* $\mathbf{fs}(Z, W)$ *is finite (hereditarily finite).*

PROOF. The proof is by mathematical induction on Z with W a fixed finite (hereditarily finite) set. Clearly, $\mathbf{fs}(0, W)$ is hereditarily finite. Assume now that $\mathbf{fs}(Z, W)$ is finite. We can introduce a function F such that $\mathbf{do}(F) = \mathbf{fs}(Z, W) \times W$, and whenever $<X, Y> \in \mathbf{do}(F)$, then $\mathbf{ap}(F, <X, Y>) = X; Y$. It follows that $\mathbf{do}(F)$ is finite, so $\mathbf{ra}(F)$ is also finite. Clearly, $\mathbf{fs}(S(Z), W) = \mathbf{ra}(F) \cup \{\emptyset\}$ is also finite. Assume now that W is hereditarily finite and $\mathbf{fs}(Z, W)$ is also hereditarily finite. Clearly, all the elements of $\mathbf{fs}(S(Z), W)$ are hereditarily finite. It follows that $\mathbf{fs}(S(Z), W)$ is hereditarily finite. □

Theorem 5.9 *Let W and X be sets and Z a natural number. The following conditions are equivalent:*

(i) $X \in \mathbf{fs}(Z, W)$.

(ii) $X \subseteq W \wedge \mathbf{FI}(X) \wedge \mathbf{ct}(X) \leq_{\mathbf{nt}} Z$.

PROOF. The implication from (i) to (ii) is proved by mathematical induction on Z with W fixed and the predicate $p(Z) \equiv (\forall X \in \mathbf{fs}(Z, W))(X \subseteq W \wedge \mathbf{FI}(Z) \wedge \mathbf{ct}(X) \leq_{\mathbf{nt}} Z)$. The case $p(0)$ is trivial. To prove $p(S(Z))$, assume $X \in \mathbf{fs}(S(Z))$. If $X = \emptyset$ we have $X \subseteq W \wedge \mathbf{FI}(X) \wedge \mathbf{ct}(X) \leq_{\mathbf{nt}} S(Z)$. If $X = X'; Y$, where $X' \in \mathbf{fs}(Z, W)$ and $Y \in W$, then by the induction hypothesis we know that $X' \subseteq W \wedge \mathbf{FI}(X') \wedge \mathbf{ct}(X') \leq_{\mathbf{nt}} Z$. It follows that $X \subseteq W \wedge \mathbf{FI}(X) \wedge \mathbf{ct}(X) \leq_{\mathbf{nt}} S(Z)$. To derive the implication from (ii) to (i) we prove by **FI**-induction on X that $X \subseteq W \rightarrow X \in \mathbf{fs}(\mathbf{ct}(X), W)$, with W fixed. The case $X = \emptyset$ is trivial, so we may assume there is $Y \in X$, and by the induction hypothesis it follows that $X \ominus Y \in \mathbf{fs}(\mathbf{ct}(X \ominus Y), W)$. It follows that $X = (X \ominus Y); Y \in \mathbf{fs}(S(\mathbf{ct}(X \ominus Y)), W)$ and $\mathbf{ct}(X) = S(\mathbf{ct}(X \ominus Y))$. Now part (i) follows by Corollary 5.7.1. □

Corollary 5.9.1 *Let W and X be arbitrary sets. The following conditions are equivalent:*

(i) $X \in \mathbf{fs}(\mathbf{ct}(X), W)$.

(ii) $X \subseteq W \wedge \mathbf{FI}(X)$.

PROOF. Immediate from Theorem 5.9 by setting $Z = \mathbf{ct}(X)$. □

Corollary 5.9.2 *Let X be an arbitrary set and W a finite set. The following conditions are equivalent:*

(i) $X \in \mathbf{fs}(\mathbf{ct}(W), W)$.

(ii) $X \subseteq W$.

PROOF. Immediate from Theorem 5.9 by setting $Z = \mathbf{ct}(W)$. □

The preceding corollary shows that whenever W is a finite set, then $\mathbf{fs}(\mathbf{ct}(W), W)$ is the set of all subsets of W (= the power-set of W).

Primitive recursion provides the basic tool for the introduction of the basic arithmetical operations: addition, multiplication, etc. We write addition with the usual in-fixed notation $+_{\mathbf{nt}}$ and multiplication with the in-fixed notation $\times_{\mathbf{nt}}$.

5.4.2-1 $X +_{\mathbf{nt}} 0 = X$.

5.4.2-2 $X +_{\mathbf{nt}} S(Z) = S(X +_{\mathbf{nt}} Z)$.

5.4.3-1 $X \times_{\mathbf{nt}} 0 = 0$.

5.4.3-2 $X \times_{\mathbf{nt}} S(Z) = (X \times_{\mathbf{nt}} Z) +_{\mathbf{nt}} X$.

In these definitions Z is intended to be a natural number and X an arbitrary set. In practice we consider only the case where X is also a natural number.

Theorem 5.10 *Assume Z and X and Y are natural numbers. Then:*

(i) *$X +_{\mathbf{nt}} Z$ is a natural number.*

(ii) $0 +_{\mathbf{nt}} Z = Z$.

(iii) $S(X) +_{\mathbf{nt}} Z = S(X +_{\mathbf{nt}} Z)$.

(iv) $X +_{\mathbf{nt}} Z = Z +_{\mathbf{nt}} X$.

(v) $(X +_{\mathbf{nt}} Y) +_{\mathbf{nt}} Z = X +_{\mathbf{nt}} (Y +_{\mathbf{nt}} Z)$.

PROOF. The five relations follow by mathematical induction on Z. Parts (i), (ii) and (iii) are trivial. In part (iv) we have $X +_{\mathbf{nt}} 0 = 0 +_{\mathbf{nt}} X$ by (ii), and $X +_{\mathbf{nt}} S(Z) = S(X +_{\mathbf{nt}} Z) = S(Z +_{\mathbf{nt}} X) = S(Z) +_{\mathbf{nt}} X$ by the definition, the induction hypothesis, and part (iii). In part (v) we have $(X +_{\mathbf{nt}} Y) +_{\mathbf{nt}} 0 = X +_{\mathbf{nt}} Y = X +_{\mathbf{nt}} (Y +_{\mathbf{nt}} 0)$. Furthermore, $(X +_{\mathbf{nt}} Y) +_{\mathbf{nt}} S(Z) = S((X +_{\mathbf{nt}} Y) +_{\mathbf{nt}} Z) = S(X +_{\mathbf{nt}} (Y +_{\mathbf{nt}} Z)) = X +_{\mathbf{nt}} S(Y +_{\mathbf{nt}} Z) = X +_{\mathbf{nt}} (Y +_{\mathbf{nt}} S(Z))$, using the induction hypothesis. □

Theorem 5.11 *Assume Z, X, and Y are natural numbers. Then:*

(i) *$X \times_{\mathbf{nt}} Z$ is a natural number.*

(ii) $0 \times_{\mathbf{nt}} Z = 0$.

(iii) $S(X) \times_{\mathbf{nt}} Z = (X \times_{\mathbf{nt}} Z) +_{\mathbf{nt}} Z$.

(iv) $X \times_{\mathbf{nt}} Z = Z \times_{\mathbf{nt}} X$.

(v) $(Y \times_{\mathbf{nt}} X) \times_{\mathbf{nt}} Z = Y \times_{\mathbf{nt}} (X \times_{\mathbf{nt}} Z)$.

(vi) $Y \times_{\mathbf{nt}} (X +_{\mathbf{nt}} Z) = (Y \times_{\mathbf{nt}} X) +_{\mathbf{nt}} (Y \times_{\mathbf{nt}} Z)$.

PROOF. Each of the six relations follows by mathematical induction on Z with X, Y fixed. Parts (i) and (ii) are trivial. In part (iii) the case $Z = 0$ is trivial. In the second case we have $S(Z) \times_{\mathbf{nt}} S(Z) = (S(X) \times_{\mathbf{nt}} Z) +_{\mathbf{nt}} S(X) = (X \times_{\mathbf{nt}} Z) +_{\mathbf{nt}} Z +_{\mathbf{nt}} S(X) = S(X \times_{\mathbf{nt}} S(Z)) +_{\mathbf{nt}} Z = (X \times_{\mathbf{nt}} S(Z)) +_{\mathbf{nt}} S(Z)$, where we use the induction hypothesis, the definition, and the associativity of addition. Parts (iv) and (v) follow easily by the induction. To prove (vi) note the case $Z = 0$ is trivial. In the second case we have $Y \times_{\mathbf{nt}} (X +_{\mathbf{nt}} S(Z)) = Y \times_{\mathbf{nt}} S(X +_{\mathbf{nt}} Z) = (Y \times_{\mathbf{nt}} (X +_{\mathbf{nt}} Z)) +_{\mathbf{nt}} Y = (Y \times_{\mathbf{nt}} X) +_{\mathbf{nt}} (Y \times_{\mathbf{nt}} Z) +_{\mathbf{nt}} Y = (Y \times_{\mathbf{nt}} X) +_{\mathbf{nt}} (Y \times_{\mathbf{nt}} S(Z))$. This completes the proof. □

EXERCISE 5.10 Assume X and Y are finite sets such that $X \cap Y = \emptyset$. Prove that $\mathbf{ct}(X \cup Y) = \mathbf{ct}(X) +_{\mathbf{nt}} \mathbf{ct}(Y)$.

5.5 Collapsing

In this section we use **WF**-recursion to define an operation **cl** (the collapsing operation) that plays an important role in modern set theory.

5.5.1 $\mathbf{cl}(Z, X) = \{\mathbf{cl}(Y, X) : Y \in Z \cap X :\}$ (*The collapsing of Z relative to X*).

Remark 5.3 This definition is intended as an application of **WF**-recursion, although we are implicitly extending the definition of **WF** to allow for one extra parameter. So we are using $\mathbf{WF}'$-induction, where $\mathbf{WF}'$ is an inductive predicate where the induction axiom takes the form,

$$\mathbf{WF}'(Z, X) \equiv (\forall Y \in Z)\mathbf{WF}'(Y, X),$$

and it follows by an easy induction that $\mathbf{WF}(Z) \equiv \mathbf{WF}'(Z, X)$ holds for arbitrary sets Z, X. Note that definition 5.5.1 implies a 3-ary set operation h, which is in fact defined in the form,

$$h(F, Z, X) = \{\mathbf{ap}(F, Y) : Y \in Z \cap X :\}.$$

The condition $Y \in Z \cap X$ is essential, because $Z \cap X \subseteq Z \subseteq \mathbf{tc}_{\mathbf{WF}'}(Z, X) = \mathbf{tc}(Z)$. For example, it would not be legitimate to use $Z \cup X$ in place of $Z \cap X$. □

EXERCISE 5.11 Prove $\mathbf{tc}_{\mathbf{WF}'}(Z, X) = \mathbf{tc}(Z)$ and $\mathbf{rk}_{\mathbf{WF}'}(Z, X) = \mathbf{rk}(Z)$ whenever Z is a well-founded set, X is an arbitrary set, and $\mathbf{WF}'$ is the predicate in Remark 5.3.

EXERCISE 5.12 Prove that if Z is well-founded and X is an arbitrary set, then $\mathbf{cl}(Z, X)$ is well-founded.

We are interested primarily in applications of the operation **cl** in the form $\mathbf{cl}(Z, Z)$, where the set Z is well-founded.

Theorem 5.12 *If Z be an arbitrary set, then:*

(i) $\mathbf{cl}(Z, Z) = \{\mathbf{cl}(Y, Z) : Y \in Z :\}$.

(ii) *The set* $\mathbf{cl}(Z, Z)$ *is transitive.*

(iii) *If $Y \in Z$, $Y' \in Z$ and $Y' \in Y$, then* $\mathbf{cl}(Y', Z) \in \mathbf{cl}(Y, Z)$.

PROOF. The three properties follow immediately from the definitions. □

We can refine the properties of the construction $\mathbf{cl}(Z, Z)$ by introducing an extra assumption in the form of a new basic predicate.

5.5.2 $\mathbf{EX}(Z) \equiv (\forall Y \in Z)(\forall Y' \in Z)(Y = Y' \vee Y \cap Z \not\subseteq Y' \vee Y' \cap Z \not\subseteq Y$ (*Z is locally extensional*).

In the next theorem we prove that whenever the set Z is well-founded and locally extensional, then the operation $\mathbf{cl}(Y, Z)$ is 1–1 when the variable Y ranges over Z.

Theorem 5.13 *Let Z be a well-founded, locally extensional set. If $Y \in Z$, $Y' \in Z$, and $\mathbf{cl}(Y, Z) = \mathbf{cl}(Y', Z)$, then $Y = Y'$.*

PROOF. With Z a fixed locally extensional set, we introduce the unary predicate p such that

$$p(Y) \equiv Y \in Z \rightarrow (\forall Y' \in Z)(\mathbf{cl}(Y, Z) = \mathbf{cl}(Y', Z) \rightarrow Y = Y'),$$

and we prove by $\mathbf{WF}$-induction that $p(Y)$ holds whenever Y is a well-founded set. The induction hypothesis is that Y is well-founded and whenever $Y'' \in Y$, then $p(Y'')$ holds. To prove $p(Y)$ we assume that $Y \in Z$ and consider $Y' \in Z$ such that $\mathbf{cl}(Y, Z) = \mathbf{cl}(Y', Z)$. To get a contradiction we assume $Y \neq Y'$, and since Z is locally extensional it follows that either $Y \cap Z \not\subseteq Y'$ or $Y' \cap Z \not\subseteq Y$. If $Y \cap Z \not\subseteq Y'$ there is $Y'' \in Y \cap Z$ and $Y'' \notin Y'$. It follows that $\mathbf{cl}(Y'', Z) \in \mathbf{cl}(Y, Z) = \mathbf{cl}(Y', Z)$, hence there is $V \in Y' \cap Z$ such that $\mathbf{cl}(Y'', Z) = \mathbf{cl}(V, Z)$. By the assumption above we have $Y'' = V$, hence $Y'' \in Y'$, which is a contradiction. The second case where $Y' \cap Z \not\subseteq Y$ is similar. □

EXERCISE 5.13 Assume the set Z is well-founded and transitive. Prove:

(a) Z is locally extensional.

(b) $\mathbf{cl}(Y, Z) = Y$ whenever $Y \in Z$.

Corollary 5.13.1 *Let Z be a well-founded, locally extensional set. If $Y \in Z$, $Y' \in Z$ and $\mathbf{cl}(Y', Z) \in \mathbf{cl}(Y, Z)$, then $Y' \in Y$.*

PROOF. From the assumption it follows that $\mathbf{cl}(Y', Z) = \mathbf{cl}(V, Z)$, where $V \in Y \cap Z$. From Theorem 5.13 it follows that $Y' = V$, hence $Y' \in Y$. □

Corollary 5.13.2 *Let Z be a well-founded, locally extensional set, $Y \in Z$, and $Y' \in Z$. Then*

(i) $\mathbf{cl}(Y', Z) = \mathbf{cl}(Y, Z) \equiv Y' = Y$.

(ii) $\mathbf{cl}(Y', Z) \in \mathbf{cl}(Y, Z) \equiv Y' \in Y$.

PROOF. Both properties follow easily from the definitions, Theorem 5.12 (iii), Theorem 5.13, and Corollary 5.13.1. □

The preceding results show that, given a locally extensional set Z, $F = (\lambda Y \in Z)\mathbf{cl}(Y, Z)$ is a 1–1 function such that $\mathbf{do}(F) = Z$, $\mathbf{ra}(F)$ is transitive, and whenever $Y \in Z$ and $Y' \in Z$, then $\mathbf{ap}(F, Y') \in \mathbf{ap}(F, Y) \equiv Y' \in Y$.

EXERCISE 5.14 Assume the set Z is well-founded and F is a function with $\mathbf{do}(F) = Z$, where $\mathbf{ra}(F)$ is transitive, that satisfies the following relation whenever $Y' \in Z$, $Y \in Z$: $Y' \in Y \equiv \mathbf{ap}(F, Y') \in \mathbf{ap}(F, Y)$. Prove that $\mathbf{ap}(F, Y) = \mathbf{cl}(Y, Z)$ whenever $Y \in Z$.

5.6 Recursion in a Local Universe

The theory of a local universe $\mathtt{U}$ is concerned primarily with predicates that are either elementary in $\mathtt{U}$ or Σ-elementary over $\mathtt{U}$. In general, inductive predicates do not belong to this class, and recursive operations are not elementary, although this may be the case in some universes.

In this section we discuss an inductive predicate ind under the assumption that the predicates q_1, q_2 and the set operation g are elementary in $\mathtt{U}$, where $\mathtt{U}$ is a fixed local universe. We say that an induction satisfying these conditions is an *elementary induction*. The theory can be extended to induction with parameters, but we do not consider this situation here.

Note that under the assumptions above it follows that the set operation $\mathbf{is}_{\mathsf{ind}}$ (3.1.1) is elementary in $\mathtt{U}$. Hence, if $Z \in \mathtt{U}$, then $\mathbf{is}_{\mathsf{ind}}(Z) \in \mathtt{U}$ and $\mathbf{is}_{\mathsf{ind}}(Z) \subseteq \mathtt{U}$.

Since the set $\mathtt{U}$ is well-founded, we can express some inductive predicates by elementary expressions. In relation to the predicate **NT** we introduce the elementary predicate $\mathbf{NT}^\circ$ as follows:

5.6.1 $\mathbf{NT}^\circ(Z) \equiv \mathbf{TR}(Z) \wedge (Z = \emptyset \vee (\mathbf{SC}(Z) \wedge (\forall Y \in Z)\mathbf{TR}(Y) \wedge (Y = \emptyset \vee \mathbf{SC}(Y))))$.

From Corollary 4.9.3 it follows that whenever Z is well-founded, then $\mathbf{NT}(Z) \equiv \mathbf{NT}^\circ(Z)$. In particular, if $F \in \mathtt{U}$ and $\mathbf{NT}^\circ(\mathbf{do}(F))$, then $\mathbf{do}(F)$ is a natural number and $\mathbf{do}(F) \in \mathtt{U}$.

To simplify the notation we introduce several auxiliary predicates.

$r_1(F) \equiv \mathbf{FU}(F) \wedge \mathbf{NT}^\circ(\mathbf{do}(F)) \wedge \mathbf{do}(F) \neq \emptyset$.

$r_2(Z, F) \equiv (\forall W \in \mathbf{do}(F))((\mathbf{do}(F) = S(W) \wedge \mathbf{ap}(F, W) \in \mathbf{is}_{\mathsf{ind}}(Z)) \vee (S(W) \in \mathbf{do}(F) \wedge \mathbf{ap}(F, W) \in \mathbf{is}_{\mathsf{ind}}(\mathbf{ap}(F, S(W)))))$.

$r(Z, F) \equiv r_1(F) \wedge r_2(Z, F)$.

These are elementary predicates, so they are independent of $\mathtt{U}$. Note that if $r(Z, F)$ holds, and $F \in \mathtt{U}$, then $\mathbf{do}(F)$ is a natural number.

EXERCISE 5.15 Assume $\mathsf{ind}(Z)$, and $r(Z, F)$, where $\mathbf{do}(F) = S(W_0)$, W_0 is a natural number, and $W_0 \neq \emptyset$. Prove $r(Y, F \restriction W_0)$ for some $Y \in \mathbf{is}_{\mathsf{ind}}(Z)$.

EXERCISE 5.16 Assume $\mathsf{ind}(Z)$ and $r(Z, F)$, where $Z \in \mathbf{is}_{\mathsf{ind}}(Z')$ and $\mathbf{do}(F)$ is a natural number. Prove that $r(Z', F \cup \{<\mathbf{do}(F), Z>\})$.

Theorem 5.14 *Assume* $\mathsf{ind}(Z)$, $r(Z, F)$, *and* $F \in \mathtt{U}$. *Then* $\mathbf{ra}(F) \subseteq \mathbf{tc}_{\mathsf{ind}}(Z)$.

PROOF. We prove by ind-induction that $\mathsf{ind}(Z) \rightarrow p(Z)$ for arbitrary set Z, where $p(Z) \equiv (\forall F \in \mathtt{U})(r(Z,F) \wedge Z \in \mathtt{U} \rightarrow \mathbf{ra}(F) \subseteq \mathbf{tc}_{\mathsf{ind}}(Z))$. We assume $F \in \mathtt{U}$, and $r(Z,F)$, to prove $\mathbf{ra}(F) \subseteq \mathbf{tc}_{\mathsf{ind}}(Z)$. We know that $\mathbf{do}(F) = S(W)$ for some natural number W, and $\mathbf{ap}(F,W) = Y$, where $Y \in \mathbf{is}_{\mathsf{ind}}(Z) \subseteq \mathbf{tc}_{\mathsf{ind}}(Z)$. If $W = \emptyset$, we are done. Otherwise, we set $F' = F{\restriction}W$, hence $F' \in \mathtt{U}$ and $r(Y,F')$ holds. By the induction hypothesis we know that $p(Y)$ holds whenever $Y \in \mathbf{is}_{\mathsf{ind}}(Z)$, hence $\mathbf{ra}(F') \subseteq \mathbf{tc}_{\mathsf{ind}}(Y) \subseteq \mathbf{tc}_{\mathsf{ind}}(Z)$. We conclude that $\mathbf{ra}(F) = \mathbf{ra}(F') \cup \{Y\} \subseteq \mathbf{tc}_{\mathsf{ind}}(Z)$. □

Now we introduce a predicate cc that is Σ-elementary over $\mathtt{U}$ and plays a crucial role in this section.

$$cc(Z,X) \equiv \mathbf{is}_{\mathsf{ind}}(Z) \subseteq X \wedge \mathbf{TR}_{\mathsf{ind}}(X) \wedge (\forall Y \in X)((\exists F \in \mathtt{U})(r(Z,F) \wedge \mathbf{ap}(F,0) = Y)).$$

EXERCISE 5.17 Prove that $cc(Z,\emptyset) \equiv \mathbf{is}_{\mathsf{ind}}(Z) = \emptyset$.

Corollary 5.14.1 *If* $\mathsf{ind}(Z)$ *and* $cc(Z,X)$, *then* $X = \mathbf{tc}_{\mathsf{ind}}(Z)$.

PROOF. From $cc(Z,X)$ we know $\mathbf{is}_{\mathsf{ind}}(Z) \subseteq X$ and $\mathbf{TR}_{\mathsf{ind}}(X)$, hence from Theorem 3.4 (ii) we have $\mathbf{tc}_{\mathsf{ind}}(Z) \subseteq X$. On the other hand, if $Y \in X$ we know there is $F \in \mathtt{U}$ and $Y \in \mathbf{ra}(F)$, so by Theorem 5.14 $Y \in \mathbf{tc}_{\mathsf{ind}}(Z)$. □

Theorem 5.15 *If* $Z \in \mathtt{U}$ *and* $\mathsf{ind}(Z)$, *then* $cc(Z, \mathbf{tc}_{\mathsf{ind}}(Z))$.

PROOF. We use ind-induction. The induction hypothesis is that whenever $Y \in \mathbf{is}_{\mathsf{ind}}(Z)$, then $cc(Y, \mathbf{tc}_{\mathsf{ind}}(Y))$. To prove $cc(Z, \mathbf{tc}_{\mathsf{ind}}(Z))$, note that $\mathbf{is}_{\mathsf{ind}}(Z) \subseteq \mathbf{tc}_{\mathsf{ind}}(Z) \wedge \mathbf{TR}_{\mathsf{ind}}(\mathbf{tc}_{\mathsf{ind}}(Z))$ follows from Theorem 3.4 (i). We assume now that $Y' \in \mathbf{tc}_{\mathsf{ind}}(Z)$. If $Y' \in \mathbf{is}_{\mathsf{ind}}(Z)$ we set $F = \langle 0, Y' \rangle$ and clearly $F \in \mathtt{U}$. If $Y' \notin \mathbf{is}_{\mathsf{ind}}(Z)$, it follows that $Y' \in \mathbf{tc}_{\mathsf{ind}}(Y)$ where $Y \in \mathbf{is}_{\mathsf{ind}}(Z)$, hence there is $F' \in \mathtt{U}$ such that $r(Y,F') \wedge \mathbf{ap}(F',0) = Y'$, so we take $F = F' \cup \{\langle \mathbf{do}(F'), Y \rangle\}$. Clearly, $F \in \mathtt{U}$ and $\mathbf{ap}(F,0) = Y'$. □

Corollary 5.15.1 *Assume* $Z \in \mathtt{U}$, $\mathsf{ind}(Z)$ *and* X *is an arbitrary set. The following conditions are equivalent:*

(i) $X = \mathbf{tc}_{\mathsf{ind}}(Z)$.

(ii) $cc(Z, X)$.

PROOF. The implication from (i) to (ii) is clear from Theorem 5.15. The implication from (ii) to (i) follows from Corollary 5.14.1. □

Theorem 5.16 *Assume* $Z \in \mathtt{U}$ *and* $\mathsf{ind}(Z)$. *Then* $\mathbf{tc}_{\mathsf{ind}}(Z) \in \mathtt{U}$.

PROOF. The proof is by ind-induction. The induction hypothesis means that whenever $Y \in \mathbf{is}_{\mathsf{ind}}(Z)$, then $\mathbf{tc}_{\mathsf{ind}}(Y) \in \mathtt{U}$. If $Z' = \mathbf{is}_{\mathsf{ind}}(Z)$ it follows from Corollary 5.15.1 that $(\forall Y \in Z')(\exists! X \in \mathtt{U})cc(Y, X)$. This means that the assumptions in the construction of Example 2.15 are satisfied, where p is the predicate cc, so there is a binary set operation f elementary in $\mathtt{U}$, and a set $W \in \mathtt{U}$, such that $\mathbf{ap}(f(Z', W), Y) = \mathbf{tc}_{\mathsf{ind}}(Y)$ whenever $Y \in Z'$. This means we can write

$$\mathbf{tc}_{\mathsf{ind}}(Z) = \mathbf{is}_{\mathsf{ind}}(Z) \cup \bigcup\{\mathbf{ap}(f(Z', W), Y) : Y \in \mathbf{is}_{\mathsf{ind}}(Z) :\}.$$

We conclude that $\mathbf{tc}_{\mathsf{ind}}(Z) \in \mathtt{U}$. □

Now we consider an arbitrary set operation rec introduced by ind-recursion in the form

$$\mathsf{rec}(Z) = h((\lambda Y \in \mathbf{tc}_{\mathsf{ind}}(Z))\mathsf{rec}(Y), Z).$$

As before, we assume the induction is elementary. We also assume the recursion is *elementary* in the sense that the set operations h is elementary in $\mathtt{U}$. This does not imply that the operation rec is elementary in $\mathtt{U}$. Still, we shall show that if $\mathsf{ind}(Z)$ and $Z \in \mathtt{U}$, then $\mathsf{rec}(Z) \in \mathtt{U}$.

We introduce a binary predicate dd that plays a role here similar to the predicate cc in relation to the operation $\mathbf{tc}_{\mathsf{ind}}$. Note that the predicate dd is Σ-elementary over $\mathtt{U}$.

$$dd(Z, F) \equiv \mathbf{FU}(F) \wedge cc(Z, \mathbf{do}(F)) \wedge (\forall Y \in \mathbf{do}(F))(\exists V \in \mathtt{U})(cc(Y, V) \wedge \mathbf{ap}(F, Y) = h((\lambda Y' \in V)\mathbf{ap}(F, Y'), Y)).$$

Remark 5.4 Assume $\mathsf{ind}(Z)$ and $Z \in \mathtt{U}$. If $dd(Z,F)$ holds we conclude from Corollary 5.15.1 that $\mathbf{do}(F) = \mathbf{tc}_{\mathsf{ind}}(Z)$. Also from Corollary 5.15.1 it follows that $(\forall Y \in \mathbf{tc}_{\mathsf{ind}}(Z))\mathbf{ap}(F,Y) = h(F',Y)$, where $F' = (\lambda Y' \in \mathbf{tc}_{\mathsf{ind}}(Y))\mathbf{ap}(F,Y')$. □

Theorem 5.17 *Assume* $\mathsf{ind}(Z)$, $Z \in \mathtt{U}$, *and* $F = (\lambda Y \in \mathbf{tc}_{\mathsf{ind}}(Z))\mathsf{rec}(Y)$. *Then* $dd(Z,X)$.

PROOF. This follows immediately from the definition. □

Theorem 5.18 *Assume* $\mathsf{ind}(Z)$, $Z \in \mathtt{U}$, *and* $dd(Z,F)$. *If* $Y \in \mathbf{tc}_{\mathsf{ind}}(Z)$, *then* $\mathbf{ap}(F,Y) = \mathsf{rec}(Y)$.

PROOF. By ind-induction on $Y \in \mathbf{tc}_{\mathsf{ind}}(Z)$ with Z and F fixed. From the induction hypothesis and Remark 5.3, it follows that whenever $Y' \in \mathbf{tc}_{\mathsf{ind}}(Y)$, then $\mathbf{ap}(F,Y') = \mathsf{rec}(Y')$. We conclude that $\mathbf{ap}(F,Y) = \mathsf{rec}(Y)$. □

Corollary 5.18.1 *Assume* $\mathsf{ind}(Z)$, $Z \in \mathtt{U}$, *and* $dd(Z,F)$. *Then* $F = (\lambda Y \in \mathbf{tc}_{\mathsf{ind}}(Z))\mathsf{rec}(Y)$ *and* $\mathsf{rec}(Z) = h(F,Z)$.

PROOF. Immediate from Theorem 5.18. □

Theorem 5.19 *Assume* $Z \in \mathtt{U}$, $\mathsf{ind}(Z)$, *and F is an arbitrary function. The following conditions are equivalent:*

(i) $F = (\lambda Y \in \mathbf{tc}_{\mathsf{ind}}(Z))\mathsf{rec}(Y)$.

(ii) $dd(Z,F)$.

PROOF. The implication from (i) to (ii) follows from Theorem 5.17. The implication from (ii) to (i) follows from Corollary 5.18.1. □

Theorem 5.20 *Assume* $\mathsf{ind}(Z)$ *and* $Z \in \mathtt{U}$. *Then* $(\lambda Y \in \mathbf{tc}_{\mathsf{ind}}(Z))\mathsf{rec}(Y) \in \mathtt{U}$.

PROOF. The proof is by ind-induction on Z. We recall that for $Z \in \mathtt{U}$ the condition $F = (\lambda Y \in \mathbf{tc}_{\mathsf{ind}}(Z))\mathsf{rec}(Y)$ is equivalent to $dd(Z, F)$. From the induction hypothesis it follows that

$$(\forall Y \in \mathbf{tc}_{\mathsf{ind}}(Z))(\exists! F_Y \in \mathtt{U})dd(Y, F_Y).$$

Note that $\mathsf{rec}(Y) = h(F_Y, Y)$ whenever $Y \in \mathbf{tc}_{\mathsf{ind}}(Z)$. Now we can apply the construction of Example 2.15 with $Z' = \mathbf{tc}_{\mathsf{ind}}(Z)$ and p the predicate dd, so it follows that there is a binary set operation f elementary in $\mathtt{U}$ and a set $W \in \mathtt{U}$ such that if $Y \in \mathbf{tc}_{\mathsf{ind}}(Z)$, then $\mathbf{ap}(f(\mathbf{tc}_{\mathsf{ind}}(Z), W), Y) = F_Y$. To simplify the notation we introduce the binary set operation f' where $f'(X, Y) = \mathbf{ap}(f(X, W), Y)$, so we have $\mathsf{rec}(Y) = h(f'(\mathbf{tc}_{\mathsf{ind}}(Z), Y), Y)$ for every $Y \in \mathbf{tc}_{\mathsf{ind}}(Z)$. We conclude that

$$F = (\lambda Y \in \mathbf{tc}_{\mathsf{ind}}(Z))h(f'(\mathbf{tc}_{\mathsf{ind}}(Z), Y), Y),$$

so $F \in \mathtt{U}$. □

Corollary 5.20.1 *If* $\mathsf{ind}(Z)$ *and* $Z \in \mathtt{U}$, *then* $\mathsf{rec}(Z) \in \mathtt{U}$.

PROOF. Immediate from the recursive definition and Theorem 5.20, noting that $\mathsf{rec}(Z) = h(F, Z)$. □

5.7 Notes

The discussion of recursion in Section 5.1 is not intended as a reduction, where a recursive evaluation is derived from more general principles, as is the case in standard set theory where the principle of transfinite recursion is actually proved from the axioms of set theory (as in von Neumann (1928), see also Levy (1979)). Rather, we take recursion as a primitive operational principle. We did something similar in Sanchis (1992), in the more restricted field of computable functionals. Still, the recursion considered there was induced by a monotonic operator, and this is not the case for set recursion, that is instead induced by set induction. The latter is not determined as the minimal fixed point of a monotonic transformation.

Our purpose in Section 5.1 is to provide a primitive construction supporting the formal axiom in the set recursion rule. Obviously, the argument here must be dependent on the discussion of the set induction rule in Chapter 3. The only requirement we impose is that the construction must be objective and independent of the universe, and these conditions are clearly satisfied in the discussion.

Several applications of recursion are given in this chapter, the intention being to show how this technique allows us to handle in quite a natural way several classical set theoretical constructions. In particular, collapsing is derived from well-founded induction via a trivial extension with an extra non-essential parameter.

While the operations of counting and power-set for finite sets are very much dependent on the particular definition of the predicate **FI**, the treatment of transitive closure and rank is a generalization of the traditional construction derived from the foundation axiom (see Levy (1979) and Drake (1974)). The collapsing operation comes from Mostowski (1949).

The material in Section 5.6 comes from Barwise (1975), where it is derived in the frame of the system **KPU**. The main result shows that a local universe is closed under set operations defined by elementary set recursion. This indicates that we cannot generate infinite sets by adding induction and recursion to the elementary rules. In the system $\mathbb{G}$ we introduce infinite sets via the omega rule in Chapter 7.

Chapter 6

Ordinals

Ordinals play a special role in set theory, due to the particular combination of global and local properties. We shall show later that the notion of ordinal is universal, for there is no set that contains all the ordinals. On the other hand, any two ordinals are identical or one of them is an element of the other. This is a local property that in general fails for well-founded sets.

We present here only the foundation of the theory of ordinals. Our approach is essentially traditional, and the reader may find extra information in any of the several expositions available in the literature.

6.1 Ordinal Induction

We define by induction the predicate **OR**. The induction axiom is as follows:

6.1.1 $\mathbf{OR}(Z) \equiv \mathbf{TR}(Z) \wedge (\forall Y \in Z)\mathbf{OR}(Y)$ (*Z is an ordinal*).

The inductive definition of **OR** is of the general form introduced in Definition 4.1.7 with $p(Z) \equiv \mathbf{TR}(Z)$, so $\mathbf{OR}(Z) \equiv \mathbf{TR}^{\dagger}(Z)$. It follows that Theorem 4.8 and Corollary 4.8.1 apply with this notation.

We shall call a proof by **OR**-induction *ordinal induction*, and a definition by **OR**-recursion we shall call *ordinal recursion*.

EXAMPLE 6.1 The preceding definition induces the following rule: If Z is transitive and all the elements in Z are ordinals, then Z is an ordinal. For example, $0 = \emptyset$ is an ordinal. The set $1 = \{0\}$ is also an ordinal. Similarly, $2 = \{0, 1\}$ is an ordinal. On the other hand, the set $\{1, 2\}$ is not an ordinal because it is not transitive. But $3 = \{0, 1, 2\}$ is an ordinal. The general rule that generates these ordinals is: $n + 1 = n \cup \{n\} = S(n)$. We conclude informally that every natural number is an ordinal. □

EXERCISE 6.1 Assume Z is an arbitrary set and prove the following conditions are equivalent:

(a) $\mathbf{OR}(Z)$.

(b) $\mathbf{WF}(Z) \wedge \mathbf{TR}(Z) \wedge (\forall Y \in \mathbf{tc}(Z))\mathbf{TR}(Y)$.

Theorem 6.1 *Assume Z is an ordinal and X is a set. Then:*

(i) *Z is well-founded and* $\mathbf{tc}_{\mathbf{OR}}(Z) = \mathbf{tc}(Z)$.

(ii) *$Z \notin Z$ and $Z \subset S(Z)$.*

(iii) *If $S(Z) = S(X)$, then $Z = X$.*

(iv) *If $X = S(Z)$, then X is an ordinal and $\bigcup X = Z$.*

(v) *If all elements in X are ordinals, then $\bigcup X$ is an ordinal.*

PROOF. Part (i) follows from Theorem 4.8 (ii), and part (ii) follows from (i) and Theorem 3.5 (i). To prove (iii), note that from (ii) and the transitivity of Z it follows that either $X \notin Z$ or $Z \notin X$, hence $Z = X$. In part (iv) we note that X is transitive and all the elements of X are ordinals, so X is an ordinal. The equality $\bigcup X = Z$ follows using the definitions and the transitivity of Z. Finally, in part (v) we note that all elements in $\bigcup X$ are ordinals. Furthermore, the union of a set of transitive sets is transitive. It follows that $\bigcup X$ is an ordinal. □

Corollary 6.1.1 *Let X be a well-founded set. The following conditions are equivalent:*

(i) *X is an ordinal.*

(ii) *X is a transitive set where every element is transitive.*

PROOF. This is a reformulation of Corollary 4.8.1 with $p(Z) \equiv \mathbf{TR}(Z)$. □

Corollary 6.1.2 *If Z be an ordinal, then $\mathbf{tc_{OR}}(Z) = \mathbf{rk_{OR}}(Z) = Z$.*

PROOF. We use ordinal induction on Z. From the induction hypothesis we get $\mathbf{tc_{OR}}(Y) = Y$ whenever $Y \in Z$. Hence,

$$\mathbf{tc_{OR}}(Z) = Z \cup \bigcup\{\mathbf{tc_{OR}}(Y) : Y \in Z :\} = Z \cup Z = Z.$$

The second part follows easily by ordinal induction using Definition 5.2.1. □

EXERCISE 6.2 Let Z be an ordinal. Prove that $\mathbf{tc}(Z) = Z$.

EXAMPLE 6.2 If Z is an ordinal and $Y \in Z$, then Y is an ordinal and $Y \subsetneq Z$. In fact, from Theorem 6.1 (ii) we get $Y \subset Z$. The converse is not true. For example, we have $\{1, 2\} \subset 3$ in Example 6.1, but $\{1, 2\} \notin 3$. On the other hand, if $Y \subset Z$ and Y is transitive, then Y is in fact an ordinal. We show later that in this case $Y \in Z$. □

Theorem 6.2 *Assume Z and Z' are ordinals. If $Y \in Z'$, then either $Z = Y$ or $Z \in Y$ or $Y \in Z$.*

PROOF. We fix Z' assuming it is an ordinal, and set the predicate p such that

$$p(Z) \equiv (\forall Y' \in Z')(Z = Y' \vee Z \in Y' \vee Y' \in Z).$$

We prove by ordinal induction that $p(Z)$ holds for every ordinal Z. With Z fixed, the induction hypothesis is that Z is an ordinal and $(\forall Y \in Z)p(Y)$.

This is called the primary induction hypothesis. We must prove that $p(Z)$ holds. Here we use ordinal induction with a predicate p' such that

$$p'(Y') \equiv Y' \notin Z' \vee Z = Y' \vee Z \in Y' \vee Y' \in Z.$$

We shall prove by ordinal induction that $p(Y')$ holds for every ordinal Y'. The induction hypothesis is that Y' is an ordinal and $(\forall V \in Y')p'(V)$. This is called the secondary induction hypothesis.

We want to prove $p'(Y')$ holds, so we assume $Y' \in Z'$, $Z \notin Y'$, $Y' \notin Z$ and show that $Z = Y'$ by extensionality.

If $V \in Z$, then by the primary induction hypothesis $p(V)$ holds. Since $V = Y'$ and $Y' \in V$ are impossible (by $V \in Z$ and $Y' \notin Z$), it follows that $V \in Y'$.

If $V \in Y'$, then by the secondary induction hypothesis $p'(V)$ holds. Since $Z = V$ and $Z \in V$ are impossible (by $V \in Y'$ and $Z \notin Y'$), it follows that $V \in Z$.

Since $p'(Y')$ holds from the secondary induction hypothesis it follows that $p'(Y')$ (with Z and Z' fixed) holds for all ordinals Y'. This means $p(Z)$ (with Z' fixed) holds from the primary induction hypothesis, hence $p(Z)$ holds for all ordinals Z. □

Corollary 6.2.1 *If Z and Y are ordinals, then one and only one of the relations, $Z = Y$, $Z \in Y$, $Y \in Z$, holds.*

PROOF. From Theorem 6.2, with $Z' = S(Y)$, it follows that one of the three relations holds. If two of the relations hold, then by transitivity we have $Z \in Z$, which contradicts Theorem 6.1 (ii). □

Remark 6.1 Corollary 6.2.1 expresses a fundamental local property in the theory of ordinals. The proof of Theorem 6.2 is by double induction, where the main induction concerns a set predicate p, and the secondary induction a set predicate p'. Note that neither of these predicates is basic, although they are general predicates covered by Theorem 3.3. □

Corollary 6.2.2 *Let X and Y be ordinals. Then:*

(i) $X \subseteq Y \vee Y \subseteq X$.

(ii) $X \nsubseteq Y \equiv Y \subset X$.

(iii) $X \not\subset Y \equiv Y \subseteq X$.

(iv) $X \in Y \equiv X \subset Y$.

(v) $S(X) \subseteq Y \equiv X \subset Y$.

(vi) $Y \subset S(X) \equiv Y \subseteq X$.

(vii) $X \in S(Y) \equiv X \subseteq Y$.

(viii) $S(X) \subseteq S(Y) \equiv X \subseteq Y$.

(ix) $S(X) \in S(Y) \equiv X \in Y$.

PROOF. Part (i) is immediate from Corollary 6.2.1 and transitivity. Parts (ii) and (iii) are immediate from (i), noting that $Y \not\subset Y$. Part (iv) from left to right follows from transitivity of Y, noting that $X \notin X$. From right to left it follows by contradiction, for if $X \notin Y$, then we have $Y \in X$ by Corollary 6.2.1, hence $Y \in Y$ from $X \subset Y$. Part (v) from left to right is trivial. From right to left it follows using part (iv). Part (vi) follows by negating both sides of (v) and using (ii) and (iii). Part (vii) follows, using (iv) and (vi). Part (viii) follows from (v) and (vi). Part (ix) follows from (iv), (v) and (vii). □

EXERCISE 6.3 Assume X and Y are ordinals. Prove:

(a) $S(X) \subset S(Y) \equiv X \subset Y$.

(b) $X \not\subset S(Y) \equiv Y \subset X$.

EXERCISE 6.4 Assume Z is an ordinal and $Y \in Z$ satisfies the condition, $(\forall V \in Z) V \subseteq Y$. Prove that $Z = S(Y)$.

EXERCISE 6.5 Let Z be an ordinal and Y a set. Prove the following conditions are equivalent:

(a) $Y \in Z$.

(b) Y is an ordinal and $Y \subset Z$.

(c) Y is transitive and $Y \subset Z$.

6.2 Infimum and Supremum

In this section we introduce set operations that are usually applied to sets where the only elements are ordinals.

6.2.1 $\mathbf{inf}(X) = \bigcap X$ (*the infimum of X*).

6.2.2 $\mathbf{sup}(X) = \bigcup X$ (*the supremum of X*).

6.2.3 $\mathbf{sup}^+(X) = \bigcup\{S(V) : V \in X :\}$ (*the strict supremum of X*).

From Theorem 6.1 (v) it follows that in case the set X contains only ordinals, then $\mathbf{sup}(X)$ and $\mathbf{sup}^+(X)$ are ordinals. For the infimum operation we show something stronger, namely, that in case X is non-empty, then $\mathbf{inf}(X) \in X$.

Theorem 6.3 *Let X be a non-empty set of ordinals, Y an arbitrary set, and $Z = \mathbf{inf}(X)$. Then:*

(i) $(\forall V \in X) Z \subseteq V$.

(ii) *If* $(\forall V \in X) Y \subseteq V$, *then* $Y \subseteq Z$.

(iii) $Z \in X \wedge Z \cap X = \emptyset$.

(iv) *Z is an ordinal.*

PROOF. Since X is non-empty, from Definition 6.2.1 it follows that the set Z satisfies the condition

$$Y \in Z \equiv (\forall V \in X) Y \in V.$$

Parts (i) and (ii) follow immediately from this condition. We prove (iii) by contradiction. Assume that $Z \neq V$ for every $V \in X$. From (i) and Corollary 6.2.1 it follows that $Z \in V$, hence $Z \in Z$, and this is a contradiction. On the other hand, if $V \in Z \cap X$, then $V \subset Z$, contradicting (i). Part (iv) follows from (iii). □

Remark 6.2 The preceding results show that $Z = \mathbf{inf}(X)$ is an ordinal whenever X is a set of ordinals. Note that either $Z = \emptyset$ or $Z \in X$, so Z is not a new ordinal. In fact, what is important in this construction is that when X is non-empty, then $Z \in X$ and $Z \subseteq V$ for every $V \in X$. □

Theorem 6.4 *Let X be a set of ordinals, Y an arbitrary set, and $Z =$* $\mathbf{sup}(X)$. *Then:*

(i) $(\forall V \in X) V \subseteq Z$.

(ii) *If* $(\forall V \in X) V \subseteq Y$, *then* $Z \subseteq Y$.

(iii) $X \subseteq S(Z)$.

(iv) $\mathbf{inf}(X) \subseteq Z$.

PROOF. From Definition 6.2.2 it follows that the set Z satisfies the condition,

$$Y \in Z \equiv (\exists V \in X) Y \in V.$$

Parts (i) and (ii) follow immediately from this condition. To prove (iii), assume $V \in X$. From (i) we have $V \subseteq Z$. If $V \subset Z$ we have $V \in Z$ by Corollary 6.2.2. Otherwise, $V = Z$, so $V \in S(Z)$. This means that $X \subseteq S(Z)$. Part (iv) is clear if X is empty. Otherwise, it follows from (i) noting $\mathbf{inf}(X) \in X$. □

EXERCISE 6.6 Let X be a set of ordinals and $Z = \mathbf{sup}(X)$. Prove that $Z \notin X \equiv X \subseteq Z$.

Remark 6.3 We now have two basic procedures to generate new ordinals from given ordinals. The first comes from Theorem 6.1 (iv), where from a given ordinal Z we can introduce $S(Z)$, which is also an ordinal. The second comes from Theorem 6.4, where if X is a set of ordinals, then $Z = \mathbf{sup}(X)$ is also an ordinal. Note that in this case it is possible that Z is an element of X. □

Theorem 6.5 *Let X be a set of ordinals, Y a set, and $Z = \mathbf{sup}^+(X)$. Then:*

(i) $Y \in Z \equiv \mathbf{OR}(Y) \wedge (\exists V \in X) Y \subseteq V$.

(ii) $X \subseteq Z$.

(iii) *If Y is an ordinal and $X \subseteq Y$, then $Z \subseteq Y$.*

(iv) $\mathbf{sup}(X) \subseteq Z \subseteq S(\mathbf{sup}(X))$.

PROOF. Part (i) follows from definition 6.2.3, noting that whenever Y and V are ordinals, then $Y \in S(V) \equiv Y \subseteq V$. Part (ii) follows from part (i) with $Y = V$. To prove (iii), assume $Y' \in Z$. From Definition 6.2.3 it follows that there is $V \in X$ such that $Y' \in S(V)$. From the assumption on Y it follows that $S(V) \subseteq Y$, hence $Y' \in Y$. To prove (iv) we note that from Theorem 6.4 (ii) it follows that $\mathbf{sup}(X) \subseteq Z$. On the other hand, if $Y \in Z$, then there is $V \in X$ such that $Y \in S(V)$, hence $Y \subseteq V \subseteq \mathbf{sup}(X)$. We conclude that $Y \in S(\mathbf{sup}(X))$. □

EXERCISE 6.7 Let X be a set of ordinals. Prove that $(\forall V \in X) S(V) \subseteq \mathbf{sup}^+(X)$ and $\mathbf{sup}^+(X) \notin X$.

Remark 6.4 We know that from part (iv) of Theorem 6.5 it follows that either $\mathbf{sup}(X) = \mathbf{sup}^+(X)$ or $\mathbf{sup}^+(X) = S(\mathbf{sup}(X))$. The next result provides more definite information about this situation. □

Corollary 6.5.1 *Let X be a set of ordinals. Then:*

(i) *If* $\mathbf{sup}(X) \in X$*, then* $\mathbf{sup}^+(X) = S(\mathbf{sup}(X))$.

(ii) *If* $\mathbf{sup}(X) \notin X$*, then* $\mathbf{sup}^+(X) = \mathbf{sup}(X)$.

PROOF. If $\mathbf{sup}(X) \in X$, then by Theorem 6.5 (ii) we have $\mathbf{sup}(X) \in \mathbf{sup}^+(X)$, hence $\mathbf{sup}^+(X) = S(\mathbf{sup}(X))$ by Theorem 6.5 (iv). On the other hand, if $\mathbf{sup}(X) \notin X$ we have $X \subseteq \mathbf{sup}(X)$ by Theorem 6.4 (v), and $\mathbf{sup}^+(X) = \mathbf{sup}(X)$ by Theorem 6.5 (iii) and (iv). □

EXERCISE 6.8 Prove that if X is an ordinal, then $\mathbf{sup}^+(X) = X$.

EXERCISE 6.9 Let X be a set of ordinals and Y a set such that $(\forall V \in X) S(V) \subseteq Y$. Prove that $\mathbf{sup}^+(X) \subseteq Y$.

Remark 6.5 From part (ii) of Theorem 6.5 it follows that $\mathbf{sup}^+(X) \notin X$. This means that we have a procedure to generate a new ordinal from a given set of ordinals. In particular, it follows that there is no set that contains all the ordinals. □

EXERCISE 6.10 Assume X is a set of ordinals and Y is a set. Prove that the following conditions are equivalent:

(a) $Y \in X \wedge (\forall V \in X) V \subseteq Y$.

(b) $Y \in X \wedge \mathbf{sup}(X) = Y$.

(c) $\mathbf{sup}^+(X) = S(Y)$.

EXERCISE 6.11 Let X be a set of ordinals. Prove that the following conditions are equivalent:

(a) $\mathbf{sup}(X) = \mathbf{sup}^+(X)$.

(b) $\mathbf{sup}(X) \notin X$.

(c) $(\forall V \in X)(\exists V' \in X) V \in V'$.

EXERCISE 6.12 Let X be a set of ordinals. Prove that the following conditions are equivalent:

(a) $\mathbf{sup}(X) \subset \mathbf{sup}^+(X)$.

(b) $\mathbf{sup}(X) \in X$.

(c) $(\exists V \in X)(\forall V' \in X) V' \subseteq V$.

6.3 Ordering

The relation $\subseteq$ is a partial order on the universe of sets, as it is reflexive and transitive. When restricted to ordinals it becomes a linear order (see Corollary 6.2.2 (i)). In fact, from Theorem 6.3 it follows that it is a local well-order.

This being the case, it is convenient to introduce a notation that emphasizes the ordering properties of the relation $\subseteq$. At the same time we note that the relation $\in$ restricted to ordinals is equivalent to $\subset$ (see Corollary 6.2.2 (iv)). For the same reasons, it is convenient to introduce a notation that makes explicit the ordering properties of the relation $\in$.

6.3.1 $X \leq_{\mathbf{or}} Y \equiv X \subseteq Y$ (*X is less or equal than Y*).

6.3.2 $X <_{\mathbf{or}} Y \equiv X \in Y$ (*X is less than Y*).

As usual, we write the negation of $\leq_{\mathbf{or}}$ in the form $\not\leq_{\mathbf{or}}$ and the negation of $<_{\mathbf{or}}$ in the form $\not<_{\mathbf{or}}$.

We intend to use the new notation only when the arguments are ordinals. As mentioned above, if X and Y are ordinals, then $X <_{\mathbf{or}} Y \equiv X \subset Y$.

With this notation it is easy to see that the results of the preceding section characterize the relations $\leq_{\mathbf{or}}$ and $<_{\mathbf{or}}$ as linear orders where the operation successor, **inf**, **sup**, and $\mathbf{sup}^+$ play special roles. We collect below all this information in a list of properties that are essentially translations of previous results. Still, in some cases the translations are valid as long as the arguments are ordinals. The reader is encouraged to verify the properties in the list below, where X, Y, and Z denote ordinals.

OR 1 $X <_{\mathbf{or}} Y \equiv X \leq_{\mathbf{or}} Y \land X \neq Y$.

OR 2 $X \leq_{\mathbf{or}} Y \equiv X <_{\mathbf{or}} Y \lor X = Y$.

OR 3 $X \not<_{\mathbf{or}} Y \equiv Y <_{\mathbf{or}} X$.

OR 4 $X <_{\mathbf{or}} S(Y) \equiv X \leq_{\mathbf{or}} Y$.

OR 5 $S(X) \leq_{\mathbf{or}} Y \equiv X <_{\mathbf{or}} Y$.

OR 6 $X \subseteq Y \equiv S(X) \subseteq S(Y)$.

OR 7 $X \in Y \equiv S(X) \in S(Y)$.

OR 8 $X <_{\mathbf{or}} Y \land Y <_{\mathbf{or}} Z \to X <_{\mathbf{or}} Z$.

OR 9 $X \leq_{\mathbf{or}} Y \land Y \leq_{\mathbf{or}} Z \to X \leq_{\mathbf{or}} Z$.

OR 10 $X <_{\mathbf{or}} Y \land Y \leq_{\mathbf{or}} Z \to X <_{\mathbf{or}} Z$.

OR 11 $X = Y \lor X <_{\mathbf{or}} Y \lor Y <_{\mathbf{or}} X$.

OR 12 $X \leq_{\mathbf{or}} Y \lor Y \leq_{\mathbf{or}} X$.

In the next items of the list, X denotes a non-empty set of ordinals, Y is a set, and Z is an ordinal.

OR 13 $(\forall V \in X)\mathbf{inf}(X) \leq_{\mathbf{or}} V$.

OR 14 If $(\forall V \in X)Y \leq_{\mathbf{or}} V$, then $Y \leq_{\mathbf{or}} \mathbf{inf}(X)$.

OR 15 $(\forall V \in X)V \leq_{\mathbf{or}} \mathbf{sup}(X)$.

OR 16 If $(\forall V \in X)V \leq_{\mathbf{or}} Y$, then $\mathbf{sup}(X) \leq_{\mathbf{or}} Y$.

OR 17 $X \subseteq \mathbf{sup}^+(X)$.

OR 18 If $X \subseteq Z$, then $\mathbf{sup}^+(X) \leq_{\mathbf{or}} Z$.

Since the notation $<_{\mathbf{or}}$ is equivalent to $\in$, it can be used to express local quantifiers. So, in place of $(\forall Y \in Z)$ we can write $(\forall Y <_{\mathbf{or}} Z)$, and in place of $(\exists Y \in Z)$ we can write $(\exists Y <_{\mathbf{or}} Z)$. In fact, this notation can be used in any situation, always with the understanding that $<_{\mathbf{or}}$ stands for $\in$. In practice, it will be used only in cases where the context implies that Z is an ordinal.

We know that $\emptyset$ is an ordinal and, furthermore, if Y is an ordinal, then $S(Y)$ is also an ordinal different from $\emptyset$. This suggests a classification of the ordinals that are different from $\emptyset$ in two types, which we formalize with the following predicates.

6.3.3 $\mathbf{SO}(X) \equiv \mathbf{OR}(X) \wedge (\exists Y \in X) X = S(Y)$ (*X is a successor ordinal*).

6.3.4 $\mathbf{LO}(X) \equiv \mathbf{OR}(X) \wedge X \neq \emptyset \wedge \neg\mathbf{SO}(X)$ (*X is a limit ordinal*).

Theorem 6.6 *If X is an ordinal, then the following conditions are equivalent:*

(i) *X is a successor ordinal.*

(ii) $\mathbf{sup}(X) <_{\mathbf{or}} X$.

(iii) $S(\mathbf{sup}(X)) = X$

PROOF. Assume (i) holds, so $X = S(Y)$ and $Y \in X$. From Theorem 6.4 it follows that $Y \leq_{\mathbf{or}} \mathbf{sup}(X) \leq_{\mathbf{or}} Y$, so $\mathbf{sup}(X) = Y$. The implication from (ii) to (iii) follows from Corollary 6.5.1, noting that X is an ordinal, so $\mathbf{sup}^+(X) = X$. The implication from (iii) to (i) is trivial. □

Corollary 6.6.1 *If X is an ordinal, then the following conditions are equivalent:*

(i) *X is a limit ordinal.*

(ii) $X \neq \emptyset \wedge \mathbf{sup}(X) = X$.

PROOF. Immediate from Theorem 6.6. □

Theorem 6.7 *Let Z be an set. The following conditions are equivalent:*

(i) *Z is a natural number.*

(ii) *Z is a finite ordinal.*

PROOF. The implication from (i) to (ii) follows easily by **NT**-induction and Theorem 4.9 (v). The implication from (ii) to (i) is proved by ordinal induction. The induction hypothesis is that whenever $Y \in Z$ and Y is finite, then Y is a natural number. Since $Y \in Z$ implies $Y \subset Z$, it follows that Y is finite, hence every element of Z is a natural number. We know that Z is transitive, so by Theorem 4.11 (iii) Z is a natural number. □

Theorem 6.8 *If Z is a limit ordinal, then Z is infinite.*

PROOF. If Z is finite, then Z is a natural number by Theorem 6.7, and this contradicts Definition 4.2.1. □

Remark 6.6 Starting with $\emptyset$ and applying the operation $S(X)$ we can generate many successor ordinals. On the other hand, there is no way to generate limit ordinals at this stage of the theory. □

In dealing with ordinals it is convenient to introduce special variables that are intended to range over ordinals. For this purpose we shall use Greek letters $\alpha, \beta, \gamma, \alpha_1$, and so on. We intend to use these symbols also in basic terms, and this requires some clarification in order to avoid ambiguities. We start by determining that, in principle, ordinal variables must be considered ordinary set variables, hence their values can be arbitrary sets. As long as an ordinal variable occurs free in a basic term there is no real difference with ordinary set variables, and the use of special symbols only means that we have the intention of (or we are particularly interested in) restricting our attention to cases where such variables are assigning ordinal values. Hence a term of the form $\alpha \in \alpha$ is meaningful whether α is an ordinal or an arbitrary set. On the other hand, if α denotes an ordinal we know the term is false.

We recall that set operations and set predicates are defined for arbitrary sets in the universe. To avoid any possible ambiguity, in some cases we shall continue defining such operations and predicates using the standard set variables (i.e., $A, B, ..., Z$), and reserve the use of ordinal variables for special applications.

On the other hand, ordinal variables will play a special role in some forms of quantifications where variables are bound to a local extension. We first discuss universal and existential quantification, where ordinal variables can be used in the form,

$$(\forall \alpha \in \mathrm{W})\mathrm{U} \equiv (\forall Y \in \mathrm{W})(\mathbf{OR}(Y) \rightarrow \mathrm{U}'),$$

$$(\exists \alpha \in \mathrm{W})\mathrm{U} \equiv (\exists Y \in \mathrm{W})(\mathbf{OR}(Y) \wedge \mathrm{U}'),$$

where U' is the result of replacing in U all the free occurrences of α with the variable Y.

Another important form of quantification takes place in applications of the rule of replacement, where we now allow abstractions of the form,

$$f(\mathbf{X}) = \{\mathrm{V} : \alpha_1 \in \mathrm{V}_1 \wedge ... \wedge \alpha_s \in \mathrm{V}_s : \mathrm{U}\},$$

where the ordinal variables $\alpha_1, ..., \alpha_s$ do not occur free in the terms $\mathrm{V}_1, ..., \mathrm{V}_s$, but may occur free in the terms V and U. This definition must be considered equivalent to the following, where the ordinal variables are replaced by ordinary set variables,

$$f(\mathbf{X}) = \{\mathrm{V}' : Y_1 \in \mathrm{V}_1 \wedge ... \wedge Y_s \in \mathrm{V}_s : \mathrm{U}' \wedge \mathbf{OR}(Y_1) \wedge ... \wedge \mathbf{OR}(Y_s)\},$$

where V' and U' are the result of replacing the free occurrences of $\alpha_1, ..., \alpha_s$ in V, U with the new ordinary variables $Y_1, ..., Y_s$.

EXAMPLE 6.3 Using replacement with ordinal bound variables we can define a unary set operation f such that

$$f(Z) = \{g(\alpha) : \alpha \in Z :\}$$

where g is a given set operation. We can eliminate the ordinal variable and write an equivalent definition with ordinary set variables,

$$f(Z) = \{g(Y) : Y \in Z : \mathbf{OR}(Y)\}.$$

If all the elements in Z are ordinals it follows that $f(Z) = g[Z]$. □

We can apply the rule of local abstraction with an ordinal variable in the form

$$F = (\lambda\alpha \in \mathrm{W})\mathrm{V},$$

where the term V may contain free occurrences of the variable α. This definition is translated into an equivalent definition where an ordinary set variable is used,

$$F = \{<Y, \mathrm{V}'> : Y \in \mathrm{W} : \mathbf{OR}(Y)\},$$

where V′ is the result of replacing the variable α in V with the variable Y. Note that the domain of F is $\mathrm{W}' = \{Y \in \mathrm{W} :: \mathbf{OR}(Y)\}$.

Let ind be an inductive predicate, possibly including parameters. We show that the rank operation associated with this predicate is closely related to ordinals.

Theorem 6.9 *If* $\mathsf{ind}(Z, \mathbf{X})$ *holds, then* $\mathbf{rk}_{\mathsf{ind}}(Z, \mathbf{X})$ *is an ordinal.*

PROOF. This follows immediately from Theorem 5.1 (i) and (iv) and Corollary 6.1.1. □

In particular, when Z is a well-founded set, then $\mathbf{rk}(Z)$ is an ordinal. Also, if $\mathbf{WF}^*(Z, R)$ holds, then $\mathbf{rk}^*(Z, R)$ is an ordinal.

Corollary 6.9.1 *If R is a relation, then $\mathbf{rk}^{**}(R)$ is an ordinal.*

PROOF. From Theorems 5.2 and 6.9. □

EXAMPLE 6.4 Consider a 1–1 function F where $\mathbf{do}(F) = X$ and $\mathbf{ra}(F) = \alpha$. We introduce a relation R such that,

$$R = \{<Y, Z> : Y \in X \wedge Z \in X : \mathbf{ap}(F, Y) \leq_{\mathbf{or}} \mathbf{ap}(F, Z)\}.$$

Clearly $\mathbf{do}(R) = \mathbf{ra}(R) = \mathbf{fd}(R) = X$. Furthermore, from Corollary 6.2.1 it follows that R is connected. We note the following consequences from the definition.

(i) If $Y \in X$ and $Z \in X$, then $Y \in \mathbf{pp}(Z, R) \equiv \mathbf{ap}(F, Y) <_{\mathbf{or}} \mathbf{ap}(F, Z)$. This follows because F is 1–1. Hence, if $Z \in X$ is R-well-ordered, then $\mathbf{pp}(Z, R) = \mathbf{tc}^*(Z, R)$ (Theorem 5.3).

(ii) Every element of X is R-well-ordered. In fact, we can prove by ordinal induction on β that whenever $\mathbf{ap}(F, Z) = \beta$, then Z is R-well-ordered. This means that R is a well-ordered relation (Definition 5.2.6).

(iii) Furthermore, we can prove that $\mathbf{rk}^*(Z, R) = \beta$ whenever $\mathbf{ap}(F, X) = \beta$. This follows by ordinal induction on β. From (ii) it follows that whenever $Y \in \mathbf{tc}^*(Z, R)$, then $\mathbf{ap}(F, Y) = \gamma$, where $\gamma <_{\mathbf{or}} \beta$, so by the induction hypothesis we have $\mathbf{rk}^*(Y, R) = \gamma$. We conclude that $\mathbf{rk}^*(Z, R) = \beta$.

(iv) From (iii) we conclude that $\mathbf{rk}^{**}(R) = \alpha$. □

EXAMPLE 6.5 Let α be an ordinal and A a set such that $A \subseteq \alpha$. Let R be the relation given by

$$R = \{<\gamma, \delta> : \gamma \in A \wedge \delta \in A : \gamma \leq_{\mathbf{or}} \delta\}.$$

Clearly, $\mathbf{fd}(R) = A$, R is connected, and every element in A is R-well-founded. □

EXERCISE 6.13 Let A and R be the sets in Example 6.5. Prove:

(a) If $\gamma \in A$, then $\mathbf{rk}^*(\gamma, R) \leq_{\mathbf{or}} \gamma$.

(b) $\mathbf{rk}^{**}(R) \leq_{\mathbf{or}} \alpha$.

EXAMPLE 6.6 Let F be a function where $A = \mathbf{do}(F)$ and $\mathbf{ra}(F) \subseteq \alpha$, and we assume $\mathbf{ra}(F)$ contains at least two different elements. We introduce a relation R such that

$$R = \{<Y, Z> : Y \in A \wedge Z \in A : \mathbf{ap}(F, Y) <_{\mathbf{or}} \mathbf{ap}(F, Z)\}.$$

Note that R is not necessarily connected. □

EXERCISE 6.14 Let F be the function in Example 6.6. Prove:

(a) $\mathbf{fd}(R) = A$.

(b) If $Z \in A$, then Z is R-well-founded.

(c) If $Z \in A$, then $\mathbf{rk}^*(Z, R) \leq_{\mathbf{or}} \mathbf{ap}(F, Z)$.

(d) $\mathbf{rk}^{**}(R) \leq_{\mathbf{or}} \alpha$.

6.4 Ordinal Recursion

From the inductive definition of the predicate **OR** we derive the corresponding rule of **OR**-recursion, or simply *ordinal recursion*. This type of recursion is called transfinite recursion in the literature. We can always write an application of ordinal recursion using the general format described in Chapter 5. Still, in many cases it is convenient to take advantage of the classification of ordinals explained in the preceding section, and write the recursion in three cases. A definition in this form is said to be by *primitive ordinal recursion*. It is, in fact, a generalization of the rule of primitive recursion over the natural numbers.

Rule of Primitive Ordinal Recursion Let g be a k-ary set operation, and h, h' be $k+2$-are set operations. We introduce a $k+1$-ary set operation rec that satisfies the following axioms, where α is an arbitrary ordinal:

POR 1 $\mathsf{rec}(0, \mathbf{X}) = g(\mathbf{X})$.

POR 2 $\mathsf{rec}(S(\alpha), \mathbf{X}) = h(\mathsf{rec}(\alpha, \mathbf{X}), \alpha, \mathbf{X})$.

POR 3 $\mathsf{rec}(\alpha, \mathbf{X}) = h'(\{\mathsf{rec}(\beta, \mathbf{X}) : \beta <_{\mathbf{or}} \alpha :\}, \alpha, \mathbf{X})$ if α is a limit ordinal.

We formalize this rule using the standard set recursion rule as follows. First, we extend **OR** with k non-essential parameters. Next, we introduce by cases a $k+3$-ary set operation r such that

$$r(\alpha, \mathbf{X}, Y_1, Y_2) = [\alpha = 0 \Rrightarrow g(\mathbf{X}), [\mathbf{SO}(\alpha) \Rrightarrow h(Y_1, \mathbf{pd}(\alpha), \mathbf{X}), h'(Y_2, \alpha, \mathbf{X})]].$$

Finally, we introduce rec by **OR**-recursion in the form,

$$\mathsf{rec}(\alpha, \mathbf{X}) = r(\alpha, \mathbf{X}, \mathsf{rec}(\mathbf{pd}(\alpha), \mathbf{X}), \{\mathsf{rec}(\beta, \mathbf{X}) : \beta <_{\mathbf{or}} \alpha :\}).$$

By using ordinal recursion we obtain operations that are intended to be meaningful when restricted to ordinal arguments. Of course, set operations are defined for arbitrary arguments in the universe. The use of ordinal recursion implies that application to non-ordinal arguments is not relevant.

A k-ary set operation f is an *ordinal operation* if whenever $f(\mathbf{X}) = Z$ and all the arguments $\mathbf{X}$ are ordinals, then Z is also an ordinal. For example, the successor operation $f(\alpha) = S(\alpha)$ is an ordinal operation. The finite union and the finite intersection operations are ordinal operations. The insertion operation is not an ordinal operation. The predecessor operation $\bigcup \alpha$ is an ordinal operation, where $\bigcup 0 = 0$, $\bigcup S(\alpha) = \alpha$ and $\bigcup \alpha = \alpha$ in case α is a limit ordinal. Note that $\bigcup X$ is well-defined for any set X.

An immediate application of ordinal recursion is the definition of the standard arithmetical operations on the ordinals. We discuss only addition and multiplication, that are extensions of the corresponding operations on the natural numbers. We use here primitive ordinal recursion. The notation for ordinal addition is the in-fixed symbol $+_{\text{or}}$, and for ordinal multiplication the notation is the in-fixed symbol $\times_{\text{or}}$. The recursion is on the variable α.

6.4.1-1 $\beta +_{\text{or}} 0 = \beta$.

6.4.1-2 $\beta +_{\text{or}} S(\alpha) = S(\beta +_{\text{or}} \alpha)$.

6.4.1-3 $\beta +_{\text{or}} \alpha = \mathbf{sup}(\{\beta +_{\text{or}} \gamma : \gamma <_{\text{or}} \alpha :\})$.

6.4.2-1 $\beta \times_{\text{or}} 0 = 0$.

6.4.2-2 $\beta \times_{\text{or}} S(\alpha) = (\beta \times_{\text{or}} \alpha) +_{\text{or}} \beta$.

6.4.2-3 $\beta \times_{\text{or}} \alpha = \mathbf{sup}(\{\beta \times_{\text{or}} \gamma : \gamma <_{\text{or}} \alpha :\})$.

Theorem 6.10 *Let α, β, and δ be arbitrary ordinals. Then:*

(i) *$\beta +_{\text{or}} \alpha$ is an ordinal.*

(ii) $\alpha +_{\text{or}} 1 = S(\alpha)$.

(iii) *If $\delta <_{\text{or}} \alpha$, then $\beta +_{\text{or}} \delta <_{\text{or}} \beta +_{\text{or}} \alpha$.*

(iv) *If $\delta \leq_{\text{or}} \beta$, then $\delta +_{\text{or}} \alpha \leq_{\text{or}} \beta +_{\text{or}} \alpha$.*

(v) $0 +_{\text{or}} \alpha = \alpha$.

(vi) *If α is a limit ordinal and $\delta <_{\text{or}} \beta +_{\text{or}} \alpha$, then $\beta +_{\text{or}} \alpha$ is a limit ordinal and there is $\gamma <_{\text{or}} \alpha$ such that $\delta <_{\text{or}} \beta +_{\text{or}} \gamma$.*

(vii) $(\beta +_{\text{or}} \delta) +_{\text{or}} \alpha = \beta +_{\text{or}} (\delta +_{\text{or}} \alpha)$.

PROOF. Part (i) follows by ordinal induction on α with β fixed, using Theorem 6.1. Part (ii) follows, noting that $1 = S(0)$. In part (iii) we use ordinal induction on α with δ and β fixed. If α is a successor ordinal, we use **OR 4**. If α is a limit ordinal, note that there is $\gamma <_{\text{or}} \alpha$ such that $\delta <_{\text{or}} \gamma$ and apply the induction hypothesis. To prove (iv), use ordinal induction on α with δ and β fixed. If α is a successor ordinal, use **OR 6**. If α is a limit ordinal, use Theorem 6.4. Part (v) follows by ordinal induction on α using Corollary 6.6.1. In part (vi) it is clear from part (iii) that $\beta +_{\mathbf{or}} \alpha$ is a limit ordinal. The second part is clear from Definition 6.4.1. To prove (vii) we use ordinal induction on α with δ and β fixed. We consider only the case where α is a limit ordinal, so by the induction hypothesis we have

$$(\beta +_{\mathbf{or}} \delta) +_{\mathbf{or}} \alpha = \mathbf{sup}(\{\beta +_{\mathbf{or}} (\delta +_{\mathbf{or}} \gamma) : \gamma <_{\mathbf{or}} \alpha :\}).$$

On the other hand, using part (vi) we have

$$\beta +_{\mathbf{or}} (\delta +_{\mathbf{or}} \alpha) = \mathbf{sup}(\{\beta +_{\mathbf{or}} \gamma' : \gamma' <_{\mathbf{or}} \delta +_{\mathbf{or}} \alpha :\}) =$$

$$\mathbf{sup}(\{\beta +_{\mathbf{or}} (\delta +_{\mathbf{or}} \gamma) : \gamma <_{\mathbf{or}} \alpha :\}).$$

This completes the proof of the theorem. □

Corollary 6.10.1 *If X is a non-empty set of ordinals and β is an ordinal, then $\beta +_{\mathbf{or}} \mathbf{sup}(X) = \mathbf{sup}(\{\beta +_{\mathbf{or}} \gamma : \gamma \in X; \})$.*

PROOF. We set $\alpha = \mathbf{sup}(X)$. If $\alpha \in X$ the relation is clear. Otherwise, α is a limit ordinal, hence $\beta +_{\mathbf{or}} \alpha = \mathbf{sup}(\{\beta +_{\mathbf{or}} \gamma : \gamma <_{\mathbf{or}} \alpha :\})$. Now if $\gamma <_{\mathbf{or}} \alpha$, then there is $\gamma' \in X$ such that $\gamma \leq_{\mathbf{or}} \gamma'$, hence $\beta +_{\mathbf{or}} \gamma \leq_{\mathbf{or}} \beta +_{\mathbf{or}} \gamma'$. On the other hand, if $\gamma \in X$, then $\gamma <_{\mathbf{or}} \alpha$. We conclude that $\beta +_{\mathbf{or}} \alpha = \mathbf{sup}(\{\beta +_{\mathbf{or}} \gamma : \gamma \in X :\})$. □

Corollary 6.10.2 *Let α be an ordinal. Then α can be expressed uniquely in the form $\alpha = \beta +_{\mathbf{or}} \mathsf{n}$ where β is not a successor ordinal and n is a natural number.*

PROOF. The proof is by ordinal induction on α. If α is not a successor ordinal, we can write $\alpha = \alpha +_{\mathbf{or}} 0$. If $\alpha = S(\delta)$, then by the induction hypothesis we have $\delta = \kappa +_{\mathbf{or}} \mathsf{n}$, hence $\alpha = S(\kappa +_{\mathbf{or}} \mathsf{n}) = \kappa +_{\mathbf{or}} S(\mathsf{n})$. To prove the uniqueness, assume $\alpha = \beta +_{\mathbf{or}} \mathsf{n} = \delta +_{\mathbf{or}} \mathsf{m}$. If $\alpha = 0$ it is clear that $\beta = \mathsf{n} = \delta = \mathsf{m} = 0$. If α is a limit ordinal, then $\mathsf{n} = \mathsf{m} = 0$, hence $\beta = \delta$. If $\alpha = S(\alpha_1)$, then $\mathsf{n} = S(\mathsf{n}_1)$ and $\mathsf{m} = S(\mathsf{m}_1)$, hence $\alpha_1 = \beta +_{\mathbf{or}} \mathsf{n}_1 = \delta +_{\mathbf{or}} \mathsf{m}_1$, so by the induction hypothesis we have $\beta = \delta$ and $\mathsf{n} = \mathsf{m}$. □

EXERCISE 6.15 Assume $\beta \leq_{\mathrm{or}} \alpha$ and $\alpha <_{\mathrm{or}} \beta +_{\mathrm{or}} \delta$. Prove there is $\delta' <_{\mathrm{or}} \delta$ such that $\alpha \leq_{\mathrm{or}} \beta +_{\mathrm{or}} \delta'$.

EXERCISE 6.16 Prove that if α and β are finite ordinals, then $\beta +_{\mathrm{or}} \alpha$ is also a finite ordinal.

EXERCISE 6.17 Let α, β, and δ be ordinals. Prove:

(a) If $\beta +_{\mathrm{or}} \delta = \beta +_{\mathrm{or}} \alpha$, then $\delta = \alpha$.

(b) If $\beta +_{\mathrm{or}} \delta <_{\mathrm{or}} \beta +_{\mathrm{or}} \alpha$, then $\delta <_{\mathrm{or}} \alpha$.

(c) If $\delta +_{\mathrm{or}} \beta <_{\mathrm{or}} \alpha +_{\mathrm{or}} \beta$, then $\delta <_{\mathrm{or}} \alpha$.

Theorem 6.11 *Let α, β, and δ be arbitrary ordinals. Then:*

(i) *$\beta \times_{\mathrm{or}} \alpha$ is an ordinal.*

(ii) $\alpha \times_{\mathrm{or}} 1 = \alpha$.

(iii) $1 \times_{\mathrm{or}} \alpha = \alpha$.

(iv) $0 \times_{\mathrm{or}} \alpha = 0$.

(v) *If $\delta <_{\mathrm{or}} \alpha$ and $0 <_{\mathrm{or}} \beta$, then $\beta \times_{\mathrm{or}} \delta <_{\mathrm{or}} \beta \times_{\mathrm{or}} \alpha$.*

(vi) $\beta \times_{\mathrm{or}} (\delta +_{\mathrm{or}} \alpha) = (\beta \times_{\mathrm{or}} \delta) +_{\mathrm{or}} (\beta \times_{\mathrm{or}} \alpha)$.

(vii) $(\beta \times_{\mathrm{or}} \delta) \times_{\mathrm{or}} \alpha = \beta \times_{\mathrm{or}} (\delta \times_{\mathrm{or}} \alpha)$.

PROOF. Part (i) follows by ordinal induction on α with β fixed. Part (ii) follows from Theorem 6.10 (v). Parts (iii) and (iv) follow by ordinal induction on α. Part (v) follows by primitive ordinal induction on α with β and δ fixed. The case $\alpha = 0$ is trivial, and in case $\delta <_{\mathrm{or}} S(\delta)$ we use **OR 4** and the induction hypothesis. In case α is a limit ordinal we note that $\beta +_{\mathrm{or}} \delta <_{\mathrm{or}} \beta +_{\mathrm{or}} \gamma <_{\mathrm{or}} \beta +_{\mathrm{or}} \alpha$, where $\gamma <_{\mathrm{or}} \alpha$. Part (vi) also follows by ordinal induction on α with β and δ fixed. The cases where $\alpha = 0$ or α is a successor ordinal are trivial. In case α is a limit ordinal, we compute

$\beta \times_{\mathbf{or}} (\delta +_{\mathbf{or}} \alpha) = \mathbf{sup}(\{\beta +_{\mathbf{or}} \gamma : \gamma <_{\mathbf{or}} \delta +_{\mathbf{or}} \alpha :\}) =$

$\mathbf{sup}(\{\beta \times_{\mathbf{or}} (\delta +_{\mathbf{or}} \gamma) : \gamma <_{\mathbf{or}} \alpha :\}) =$

$\mathbf{sup}(\{(\beta \times_{\mathbf{or}} \delta) +_{\mathbf{or}} (\beta \times_{\mathbf{or}} \gamma) : \gamma <_{\mathbf{or}} \alpha :\}) =$

$(\beta \times_{\mathbf{or}} \delta) +_{\mathbf{or}} (\mathbf{sup}(\{\beta \times_{\mathbf{or}} \gamma : \gamma <_{\mathbf{or}} \alpha :\}) = (\beta \times_{\mathbf{or}} \delta) +_{\mathbf{or}} (\beta \times_{\mathbf{or}} \alpha).$

The proof of part (vii) is by ordinal induction on α with δ and β fixed, similar to the proof of Theorem 6.10 (vii). □

EXERCISE 6.18 Let X be a non-empty set of ordinals and β an ordinal. Prove that $\beta \times_{\mathbf{or}} \mathbf{sup}(X) = \mathbf{sup}(\{\beta \times_{\mathbf{or}} \gamma : \gamma \in X :\})$.

EXERCISE 6.19 Prove that if α and β are finite ordinals, then $\beta \times_{\mathbf{or}} \alpha$ is also a finite ordinal.

EXERCISE 6.20 Let α, β, and δ be ordinals. Prove:

(a) If $\beta \times_{\mathbf{or}} \delta = \beta \times_{\mathbf{or}} \alpha$, then $\delta = \alpha$.

(b) If $\beta \times_{\mathbf{or}} \delta <_{\mathbf{or}} \beta \times_{\mathbf{or}} \alpha$, then $\delta <_{\mathbf{or}} \alpha$.

(c) If $\delta \leq_{\mathbf{or}} \beta$, then $\delta +_{\mathbf{or}} \alpha \leq_{\mathbf{or}} \beta +_{\mathbf{or}} \alpha$.

(d) If $\beta \times_{\mathbf{or}} \delta <_{\mathbf{or}} \beta \times_{\mathbf{or}} \alpha$, then $\delta <_{\mathbf{or}} \alpha$.

6.5 Bounded Minimalization

Minimalization is provided essentially by the operation **inf** via Theorem 6.3. Here it is formalized as an operation rule that induces a new term construction.

Rule of Bounded Minimalization Let p be a $k+1$-ary predicate. We introduce a $k+1$-ary set operation h such that

$$h(Z, \mathbf{X}) = \mathbf{inf}(\{Y \in Z :: p(Y, \mathbf{X})\}).$$

We say that h is *defined by bounded minimalization from the predicate* p. Note that in case the condition $(\exists\alpha \in Z)p(\alpha, \mathbf{X})$ fails, the value of $h(Z, \mathbf{X})$ is 0.

EXAMPLE 6.7 If $p(\alpha, Z) \equiv \alpha \in Z$, then $h(Z)$ denotes the least ordinal in Z, and in case there is no ordinal in Z, then $h(Z) = 0$. □

Remark 6.7 The rule of bounded minimalization is not a primitive rule in our system. It is obtained by substitution, via Theorem 2.1, from previously defined operations. □

Theorem 6.12 *Let h be defined by bounded minimalization from the predicate p and assume $h(Z, \mathbf{X}) = V$, where Z is a set of ordinals. Then*

(i) $\mathbf{OR}(V)$.

(ii) $\alpha \in Z \wedge p(\alpha, \mathbf{X}) \rightarrow V \in Z \wedge p(V, \mathbf{X}) \wedge V \leq_{\mathbf{or}} \alpha$.

(iii) $(\forall\alpha \in Z)\neg p(\alpha, \mathbf{X}) \rightarrow V = 0$.

(iv) $V \in Z \equiv (\exists\alpha \in Z)p(\alpha, \mathbf{X}) \vee 0 \in Z$.

PROOF. Parts (i), (ii), and (iii) are clear from the definition. To prove (iv), we use the definition and (ii). □

In applications it is more convenient to use the rule of bounded minimalization with a boolean basic term rather than a predicate. Furthermore, the argument Z of the rule can be assumed to come from a set term that depends on the input variables. In this way we obtain a new rule for the construction of basic terms.

BT 10 If U is a boolean term, W is a set term, and the variable α does not occur free in W, then $(\mu\alpha \in \mathrm{W})\mathrm{U}$ is also a set term.

The semantics for the new term is derived by setting a predicate p such that $p(Z, \mathbf{X}) \equiv \mathrm{U}'$, where U' is obtained from U by the replacement of the variable α with a new ordinary variable Z, defining h from p by bounded minimalization, and substituting W for Z.

EXERCISE 6.21 Evaluate the expression: $(\mu\alpha \in \{\emptyset\})\alpha \not\subseteq \{\emptyset\}$.

The rule of bounded minimalization provides the right tool to define the inverse of ordinal addition (ordinal subtraction) and the inverse of ordinal multiplication (ordinal division).

6.5.1 $\alpha -_{\mathbf{or}} \beta = (\mu\delta \leq_{\mathbf{or}} \alpha)\alpha \leq_{\mathbf{or}} \beta +_{\mathbf{or}} \delta$ *(the difference of α and β).*

6.5.2 $\alpha \div_{\mathbf{or}} \beta = \bigcup(\mu\delta \leq_{\mathbf{or}} S(\alpha))\alpha <_{\mathbf{or}} \beta \times_{\mathbf{or}} \delta$ *(the quotient of α by β).*

Note that $\delta \leq_{\mathbf{or}} \alpha$ is equivalent to $\delta \in S(\alpha)$. Furthermore, the operation $-_{\mathbf{or}}$ is well-defined, because $\delta = \alpha$ satisfies the condition $\alpha \leq_{\mathbf{or}} \beta +_{\mathbf{or}} \delta$. On the other hand the operation $\alpha \div_{\mathbf{or}} \beta$ is not well-defined when $\beta = 0$, hence $\alpha \div_{\mathbf{or}} 0 = \alpha$. If $0 <_{\mathbf{or}} \beta$, the operation $\alpha \div_{\mathbf{or}} \beta$ is well-defined, because $\delta = S(\alpha)$ satisfies the condition $\alpha <_{\mathbf{or}} \beta \times_{\mathbf{or}} \delta$.

Theorem 6.13 *Let α and β be ordinals where $\beta \leq_{\mathbf{or}} \alpha$. Then:*

(i) $\beta +_{\mathbf{or}} (\alpha -_{\mathbf{or}} \beta) = \alpha$.

(ii) *If $\beta +_{\mathbf{or}} \delta = \alpha$, then $\delta = \alpha -_{\mathbf{or}} \beta$.*

PROOF. We know that $\alpha -_{\mathbf{or}} \beta$ is the least ordinal δ that satisfies the condition $\alpha \leq_{\mathbf{or}} \beta +_{\mathbf{or}} \delta$. If $\alpha <_{\mathbf{or}} \beta +_{\mathbf{or}} \delta$, then $0 <_{\mathbf{or}} \delta$ because $\beta \leq_{\mathbf{or}} \alpha$ and there is $\delta' <_{\mathbf{or}} \delta$ such that $\alpha \leq_{\mathbf{or}} \beta +_{\mathbf{or}} \delta'$, contradicting the minimality of δ. We conclude that $\alpha = \beta +_{\mathbf{or}} (\alpha -_{\mathbf{or}} \beta)$. Part (ii) is clear because if $\alpha -_{\mathbf{or}} \beta <_{\mathbf{or}} \delta$, then $\alpha <_{\mathbf{or}} \beta +_{\mathbf{or}} \delta = \alpha$, and this is a contradiction. □

EXERCISE 6.22 Let α and β be arbitrary ordinals. Prove that $\alpha \leq_{\mathbf{or}} \beta$ if and only if $\alpha -_{\mathbf{or}} \beta = 0$.

EXERCISE 6.23 Assume $\beta <_{\mathrm{or}} \beta' \leq_{\mathrm{or}} \alpha$. Prove that $\alpha -_{\mathrm{or}} \beta' \leq_{\mathrm{or}} \alpha -_{\mathrm{or}} \beta$.

Theorem 6.14 *If α and β are ordinals, where $0 <_{\mathrm{or}} \beta$, then:*

(i) $S(\alpha \div_{\mathrm{or}} \beta) = (\mu\delta <_{\mathrm{or}} \alpha)\alpha <_{\mathrm{or}} (\beta \times_{\mathrm{or}} \delta)$.

(ii) $\alpha <_{\mathrm{or}} \beta \times_{\mathrm{or}} S(\alpha \div_{\mathrm{or}} \beta)$.

(iii) $\beta \times_{\mathrm{or}} (\alpha \div_{\mathrm{or}} \beta) \leq_{\mathrm{or}} \alpha$.

(iv) $\alpha = (\beta \times_{\mathrm{or}} (\alpha \div_{\mathrm{or}} \beta)) +_{\mathrm{or}} (\alpha -_{\mathrm{or}} (\beta \times_{\mathrm{or}} (\alpha \div_{\mathrm{or}} \beta)))$.

(v) $\alpha -_{\mathrm{or}} (\beta \times_{\mathrm{or}} (\alpha \div_{\mathrm{or}} \beta)) <_{\mathrm{or}} \beta$.

PROOF. Let δ be the least ordinal such that $\alpha <_{\mathrm{or}} (\beta \times_{\mathrm{or}} \delta)$. Clearly, $0 <_{\mathrm{or}} \delta$. Furthermore, δ is not a limit ordinal, for in such a case there is $\delta' <_{\mathrm{or}} \delta$ such that $\alpha <_{\mathrm{or}} (\beta \times_{\mathrm{or}} \delta')$, contradicting the minimality of δ. It follows that δ is a successor ordinal, so $\delta = S(\alpha \div \beta)$ and $\alpha <_{\mathrm{or}} \beta \times_{\mathrm{or}} S(\alpha \div \beta)$. This proves (i) and (ii). Part (iii) follows because $\alpha <_{\mathrm{or}} \beta \times_{\mathrm{or}} (\alpha \div_{\mathrm{or}} \beta)$ contradicts the minimality of $\delta = S(\alpha \div_{\mathrm{or}} \beta)$. Part (iv) follows from (iii) and Theorem 6.13 (i). Finally, we get (v), noting that if $\beta \leq_{\mathrm{or}} \alpha -_{\mathrm{or}} (\beta \times_{\mathrm{or}} (\alpha \div_{\mathrm{or}} \beta))$, then we get, using (iv), that

$$\beta \times_{\mathrm{or}} S(\alpha \div_{\mathrm{or}} \beta) = (\beta \times_{\mathrm{or}} (\alpha \div_{\mathrm{or}} \beta)) +_{\mathrm{or}} \beta \leq_{\mathrm{or}} \alpha,$$

contradicting part (i). □

Corollary 6.14.1 *Let α, β, δ and γ be ordinals where $0 \leq_{\mathrm{or}} \gamma <_{\mathrm{or}} \beta$ and $\alpha = (\beta \times_{\mathrm{or}} \delta) +_{\mathrm{or}} \gamma$. Then, $\delta = \alpha \div_{\mathrm{or}} \beta$ and $\gamma = \alpha -_{\mathrm{or}} (\beta \times_{\mathrm{or}} (\alpha \div_{\mathrm{or}} \beta))$.*

PROOF. We have $\alpha = (\beta \times_{\mathrm{or}} \delta) +_{\mathrm{or}} \gamma <_{\mathrm{or}} (\beta \times_{\mathrm{or}} \delta) +_{\mathrm{or}} \beta = \beta \times_{\mathrm{or}} S(\delta)$. From the minimality of $S(\alpha \div_{\mathrm{or}} \beta)$ it follows that $S(\alpha \div_{\mathrm{or}} \beta) \leq_{\mathrm{or}} S(\delta)$. To get a contradiction, assume $S(\alpha \div_{\mathrm{or}} \beta) <_{\mathrm{or}} S(\delta)$, hence $S(\alpha \div_{\mathrm{or}} \beta) \leq_{\mathrm{or}} \delta$. We conclude that $\alpha <_{\mathrm{or}} \beta \times_{\mathrm{or}} S(\alpha \div_{\mathrm{or}} \beta) \leq_{\mathrm{or}} \beta \times_{\mathrm{or}} \delta \leq_{\mathrm{or}} \alpha$, and this is a contradiction. It follows that $S(\alpha \div_{\mathrm{or}} \beta) = S(\delta)$, hence $\alpha \div_{\mathrm{or}} \beta = \delta$. From this and Theorem 6.13 it follows that $\alpha -_{\mathrm{or}} (\beta \times_{\mathrm{or}} (\alpha \div_{\mathrm{or}} \beta)) = \gamma$. □

6.6 Ordinal Counting

We can use bounded minimalization inside a recursive definition, and this procedure usually makes the definition more transparent. As an example we introduce a binary ordinal counting operation that involves minimalization in the following way:

6.6.1 $\mathbf{oc}(\alpha, X) = (\mu\beta \in X)(\forall\gamma <_{\mathbf{or}} \alpha)\mathbf{oc}(\gamma, X) <_{\mathbf{or}} \beta$ *(the αth ordinal in X).*

Remark 6.8 We are interested in applications of **oc** where the first argument is an ordinal and the second argument is an arbitrary set. The operation $\mathbf{oc}(\alpha, X)$ enumerates the ordinals in the set X, hence $\mathbf{oc}(0, X)$ is the smallest ordinal in X, $\mathbf{oc}(1, X)$ is the next, etc. If X contains no ordinal, then $\mathbf{oc}(\alpha, X) = 0$ for every α. If $X = \beta$ is an ordinal, then $\mathbf{oc}(\gamma, \beta) = \gamma$ if $\gamma <_{\mathbf{or}} \beta$ and $\mathbf{oc}(\gamma, \beta) = \beta$ if $\beta \leq_{\mathbf{or}} \gamma$. □

Theorem 6.15 *Let α be an ordinal and X an arbitrary set. Then:*

(i) *If $\mathbf{oc}(\alpha, X) = V$, then V is an ordinal.*

(ii) $\mathbf{oc}(\alpha, X) = (\mu\beta \in X)(\forall\gamma <_{\mathbf{or}} \alpha)\mathbf{oc}(\gamma, X) <_{\mathbf{or}} \beta$.

(iii) *If $\mathbf{oc}(\alpha, X) = \beta$, where $\beta \in X$ and $\gamma <_{\mathbf{or}} \alpha$, then $\mathbf{oc}(\gamma, X) = \delta$, where $\delta \in X$ and $\delta <_{\mathbf{or}} \beta$.*

(iv) *Let β be an ordinal in X such that whenever $\gamma <_{\mathbf{or}} \alpha$, then $\mathbf{oc}(\gamma, X) <_{\mathbf{or}} \beta$. Then $\mathbf{oc}(\alpha, X) \leq_{\mathbf{or}} \beta$.*

PROOF. Part (i) is clear from the definition of bounded minimalization, and part (ii) follows from the definition of **oc**, noting that by (i) $\mathbf{oc}(\gamma, X)$ is an ordinal. To prove (iii), note that $\delta <_{\mathbf{or}} \beta$ is clear from (ii) and the definition of unbounded minimalization. We have also $\gamma \in X$, for otherwise $\delta = \mathbf{sup}^{+}(X)$, and this contradicts $\delta <_{\mathbf{or}} \beta$. Finally, part (iv) expresses the fact that $\mathbf{oc}(\alpha, X)$ is the least ordinal that satisfies the condition. □

Theorem 6.16. *Assume $\beta \in X$. There is an ordinal $\alpha \leq_{\mathbf{or}} \beta$ such that* $\mathbf{oc}(\alpha, X) = \beta$.

PROOF. Note that from Theorem 6.15 (iii) the α is unique. The proof is by ordinal induction on β . The induction hypothesis is that whenever $\delta \in \beta$ and $\delta \in X$ there is $\gamma \leq_{\mathbf{or}} \delta$ such that $\mathbf{oc}(\gamma, X) = \delta$. Using replacement, we introduce the set Z such that

$$Z = \{\gamma \in \beta :: (\exists \delta \in \beta)\mathbf{oc}(\gamma, X) = \delta\}.$$

We note that the set Z is transitive, for if $\gamma \in Z$, $\mathbf{oc}(\gamma, X) = \delta$, and if $\gamma' \in \gamma$, then $\mathbf{oc}(\gamma', X) = \delta'$ and $\delta' <_{\mathbf{or}} \delta <_{\mathbf{or}} \beta$ by Theorem 6.15 (iii). This means that $\gamma' \in Z$. If Z is transitive, then Z is an ordinal, for Z contains only ordinals. We set $\alpha = Z$, and clearly $\alpha \leq_{\mathbf{or}} \beta$. Now we show that $\mathbf{oc}(\alpha, X) = \beta$. In fact, from Theorem 6.15 (iv) it follows that $\mathbf{oc}(\alpha, X) \leq_{\mathbf{or}} \beta$. On the other hand, if $\mathbf{oc}(\alpha, X) <_{\mathbf{or}} \beta$, then from the induction hypothesis it follows that $\alpha \in \alpha$, and this is a contradiction. We conclude that $\mathbf{oc}(\alpha, X) = \beta$. □

We are now in position to identify the upper bound for all ordinals that are required to enumerate the ordinals in a set X. We introduce the unary set operation $\mathbf{oc}'$ such that:

6.6.2 $\mathbf{oc}'(X) = \{\alpha \in \mathbf{sup}^+(X) :: \mathbf{oc}(\alpha, X) \in X\}$ (*the ordinal length of X*).

Clearly, $\mathbf{oc}'(X)$ is transitive, so it is an ordinal. Furthermore, an ordinal β is an element of X if and only if there is $\alpha <_{\mathbf{or}} \mathbf{oc}'(X)$ and $\mathbf{oc}(\alpha, X) = \beta$.

EXERCISE 6.24 Let F be a function such that $\mathbf{do}(F)$ is an ordinal and whenever $\gamma <_{\mathbf{or}} \alpha <_{\mathbf{or}} \mathbf{do}(F)$, then $\mathbf{ap}(F, \gamma) <_{\mathbf{or}} \mathbf{ap}(F, \alpha)$. Prove that $X = \mathbf{ra}(F)$ is a set of ordinals, and $F = (\lambda\alpha \in \mathbf{oc}'(X))\mathbf{oc}(\alpha, X)$.

6.7 A Pairing Operation

We have already a general pairing operation, where two sets X and Y can be encoded in the set $<X,Y>$. We introduce here a pairing operation for ordinals, where two ordinals α and β can be encoded in the ordinal $\ll\alpha,\beta\gg$. This operation will be applied later in this work.

The pairing operation depends on an auxiliary set operation g defined by primitive ordinal recursion as follows:

$g(0) = 0.$

$g(S(\alpha)) = S((g(\alpha) +_{\mathsf{or}} g(\alpha)).$

$g(\alpha) = \mathbf{sup}(\{g(\gamma) : \gamma <_{\mathbf{or}} \alpha :\})$ (α a limit ordinal).

From the definition it is clear that $g(\alpha)$ is an ordinal for every ordinal α. Furthermore, if α is a natural number (= a finite ordinal), then $g(\alpha)$ is also a natural number. For example, $g(1) = 1$, $g(2) = 3$, $g(3) = 7$, etc.

Theorem 6.17 *Let α and β be arbitrary ordinals. Then:*

(i) $\alpha \leq_{\mathbf{or}} g(\alpha)$.

(ii) $g(\alpha) +_{\mathsf{or}} g(\alpha) <_{\mathbf{or}} g(S(\alpha))$.

(iii) *If $\beta <_{\mathbf{or}} \alpha$, then $g(\beta) \leq_{\mathbf{or}} g(\alpha)$*

PROOF. Part (i) follows by ordinal induction on α. The three cases are trivial. Part (ii) follows from the definition of $g(S(\alpha))$. Part (iii) follows by ordinal induction on α with β fixed. The case $\alpha = 0$ is trivial. If $\beta <_{\mathbf{or}} S(\alpha)$, then $\beta \leq_{\mathbf{or}} \alpha$ and from the induction hypothesis we have $g(\beta) \leq_{\mathbf{or}} g(\alpha) <_{\mathbf{or}} g(S(\alpha))$. The case in which α is a limit ordinal follows from the definitions and the induction hypothesis. □

Remark 6.9 We refer to g as the gap operation, because for every α the gap between $g(\alpha)$ and $g(S(\alpha))$ is at least $g(\alpha)$. In fact, we have $g(\alpha) <_{\mathbf{or}} g(\alpha) +_{\mathsf{or}} 1 <_{\mathbf{or}} \dots <_{\mathbf{or}} g(\alpha) +_{\mathsf{or}} g(\alpha) <_{\mathbf{or}} g(S(\alpha))$, hence $g(\alpha) <_{\mathbf{or}} g(S(\alpha)) -_{\mathbf{or}} g(\alpha)$. □

We use the gap operation g to define the pairing operation.

6.7.1 $\ll\alpha, \beta\gg = g(\alpha +_{\text{or}} \beta) +_{\text{or}} g(\alpha)$.

Clearly, $\ll\alpha, \beta\gg$ is always an ordinal. If α and β are natural numbers we can show that $\ll\alpha, \beta\gg$ is also a natural number. For example, $\ll 1, 1\gg = g(2) +_{\mathbf{nt}} g(1) = 3 +_{\mathbf{nt}} 1 = 4$.

We must show that this pairing operation is 1–1, hence if $\ll\alpha, \beta\gg = \ll\alpha', \beta'\gg$, then $\alpha = \alpha'$ and $\beta = \beta'$. The best approach is to define the inverse (decoding) operations. This requires an auxiliary operation $g^{\ddagger}$ such that

$$g^{\ddagger}(\gamma) = (\mu\delta \leq_{\text{or}} \gamma)(g(\delta) \leq_{\text{or}} \gamma <_{\text{or}} g(S(\delta))).$$

Clearly, if γ is a natural number, then $g^{\ddagger}(\gamma)$ is also a natural number. The evaluation of $g^{\ddagger}(\gamma)$ is very much facilitated by the next results.

Theorem 6.18 *Let γ and δ be ordinals that satisfy the condition, $g(\delta) \leq_{\text{or}} \gamma <_{\text{or}} g(S(\delta))$. Then $g^{\ddagger}(\gamma) = \delta$.*

PROOF. Since $\delta \leq_{\text{or}} g(\delta) \leq_{\text{or}} \gamma$, it follows that δ satisfies the condition in the definition of $g^{\ddagger}(\gamma)$. Assume that $\delta_1 <_{\text{or}} \delta$ is another ordinal such that $g(\delta_1) \leq_{\text{or}} \gamma <_{\text{or}} g(S(\delta_1))$. It follows that $S(\delta_1) \leq_{\text{or}} \delta$, hence $\gamma <_{\text{or}} g(S(\delta_1)) \leq_{\text{or}} g(\delta) \leq_{\text{or}} \gamma$, and this is a contradiction. We conclude that $g^{\ddagger}(\gamma) = \delta$. □

Theorem 6.19 *If α and γ are ordinals such that $\gamma \leq_{\text{or}} g(\alpha)$, then $g^{\ddagger}(g(\alpha) +_{\text{or}} \gamma) = \alpha$.*

PROOF. From the gap conditions it follows that

$$g(\alpha) \leq_{\text{or}} g(\alpha) +_{\text{or}} \gamma \leq_{\text{or}} g(\alpha) +_{\text{or}} g(\alpha) <_{\text{or}} g(S(\alpha)),$$

hence by Theorem 6.18 we have $g^{\ddagger}(g(\alpha) +_{\text{or}} \gamma) = \alpha$. □

Corollary 6.19.1 $g^{\ddagger}(g(\alpha)) = \alpha$.

PROOF. From Theorem 6.19 with $\gamma = 0$. $\square$

Corollary 6.19.2 *If α and β are arbitrary ordinals, then $g^{\ddagger}(g(\alpha +_{\mathrm{or}} \beta) +_{\mathrm{or}} g(\alpha)) = \alpha +_{\mathrm{or}} \beta$.*

PROOF. From Theorem 6.19 with $\gamma = g(\alpha) \leq_{\mathrm{or}} g(\alpha +_{\mathrm{or}} \beta)$. $\square$

We are now in position to define the inverses of the pairing operation.

6.7.2 $\ltimes(\gamma) = g^{\ddagger}(\gamma -_{\mathrm{or}} g(g^{\ddagger}(\gamma)))$.

6.7.3 $\rtimes(\gamma) = g^{\ddagger}(\gamma) -_{\mathrm{or}} \ltimes(\gamma)$.

For example, $\ltimes(5) = 1$ and $\rtimes(5) = 1$. Also, $\ltimes(4) = 1$ and $\rtimes(4) = 1$. Recall that $\ll 1, 1 \gg = 4$.

Theorem 6.20 *Let α and β be arbitrary ordinals. Then:*

(i) $\ltimes(\ll\alpha, \beta\gg) = \alpha$.

(ii) $\rtimes(\ll\alpha, \beta\gg) = \beta$.

PROOF. From Corollary 6.19.2 it follows that $g(g^{\ddagger}(\ll\alpha, \beta\gg)) = g(\alpha +_{\mathrm{or}} \beta)$. Hence $\rtimes(\ll\alpha, \beta\gg) = g^{\ddagger}((g(\alpha +_{\mathrm{or}} \beta) +_{\mathrm{or}} g(\alpha)) -_{\mathrm{or}} g(\alpha +_{\mathrm{or}} \beta)) = g^{\ddagger}(g(\alpha)) = \alpha$. The proof of (ii) is similar. $\square$

Corollary 6.20.1 *If $\ll\alpha, \beta\gg = \ll\alpha', \beta'\gg$, then $\alpha = \alpha'$ and $\beta = \beta'$.*

PROOF. Immediate from Theorem 6.20. $\square$

EXERCISE 6.25 Let γ be an ordinal. Prove that $\ltimes(\gamma) \leq_{\mathrm{or}} \gamma$ and $\rtimes(\gamma) \leq_{\mathrm{or}} \gamma$.

6.7.4 $\mathbf{CR}(\kappa) \equiv (\forall\alpha <_{\mathrm{or}} \kappa)(\forall\beta <_{\mathrm{or}} \kappa)\ll\alpha, \beta\gg <_{\mathrm{or}} \kappa$ (*κ is a critical ordinal*).

For example, $g(0) = 0$, hence $\ll 0, 0 \gg = 0$, so $1 = \{0\}$ is a critical ordinal. On the other hand, we can see informally that if κ is a finite ordinal such that $1 <_{\text{or}} \kappa$, then κ is not a critical ordinal, no matter how the pairing is defined, because the number of pairs in $\kappa \times \kappa$ is greater than κ. So, critical ordinals must be infinite ordinals, and at this stage we have no example of such sets.

As usual (see Section 2.5), $\mathtt{U}$ denotes a fixed local universe.

6.7.5 $\kappa_{\mathtt{U}} = \mathbf{sup}^{+}(\{\gamma : \gamma \in \mathtt{U} :\})$ *(the least ordinal not in* $\mathtt{U}$*).*

Theorem 6.21 *$\kappa_{\mathtt{U}}$ is a critical ordinal.*

PROOF. From Corollary 5.20.1 we know that $\mathtt{U}$ is closed under the ordinal operation addition and also under the operation g. Since the operation 6.7.1 is defined by substitution with g and ordinal addition, it follows that $\mathtt{U}$ is closed under the pairing operation. Hence, if $\alpha <_{\text{or}} \kappa_{\mathtt{U}}$ and $\beta <_{\text{or}} \kappa_{\mathtt{U}}$, it follows that $\alpha \in \mathtt{U}$ and $\beta \in \mathtt{U}$, so $\ll \alpha, \beta \gg \in \mathtt{U}$. We conclude that $\ll \alpha, \beta \gg <_{\text{or}} \kappa_{\mathtt{U}}$. □

Corollary 6.21.1 *Let X be a non-empty set where all the elements are critical ordinals. Then, $\bigcup X$ is a critical ordinal.*

PROOF. Assume $\kappa' = \bigcup X$, and $\kappa' \notin X$. If $\alpha \leq_{\text{or}} \beta <_{\text{or}} \kappa'$, then there is $\kappa'' \in X$ such that $\beta <_{\text{or}} \kappa''$, hence $\ll \alpha, \beta \gg <_{\text{or}} \kappa'' <_{\text{or}} \kappa'$. □

EXAMPLE 6.8 A critical ordinal κ induces a partial reduction of $\kappa \times \kappa$ to κ. By this we mean a function G defined as

$$G = \{<<\alpha, \beta>, \ll \alpha, \beta \gg> : \alpha <_{\text{or}} \kappa \wedge \beta <_{\text{or}} \kappa :\}.$$

Clearly, $\mathbf{do}(G) = \kappa \times \kappa$. From Corollary 6.20.1 it follows that G is 1–1. Furthermore, since κ is critical, it follows that $\mathbf{ra}(G) \subseteq \kappa$. In Example 7.3 we show how to extend G to a total reduction G' where $\mathbf{ra}(G') = \kappa$. □

6.8 Notes

There are several explicit definitions of ordinals in standard set theory, but for a number of reasons they are not available to us. In some cases the axiom of foundation is required, which is not a part of our theory. In others, quantification over the power-set is used, which is not available at this stage of the theory. Still, it is possible to define ordinals explicitly from the predicate **WF** (see Corollary 6.1.1), but we prefer to give a primitive inductive definition from which we can immediately derive rules of proof by ordinal induction and definition by ordinal recursion. The idea of Definition 6.1.1 comes from Barwise (1975).

Leaving aside technical details, we approach ordinals in what is now the traditional way, which originates with von Neumann (1923), and there is very little that is new in this chapter. The most important result is the crucial Theorem 6.2, which requires a double induction.

Ordinals play a singular role in our theory, well beyond the central role they play in standard set theory. The whole idea of the permutation rule introduced in Chapter 8 depends on the notion of ordinal, and from such rule we derive the general power-set construction and the theory of cardinals.

Several pairing operations are available in the literature. The one we use in Section 6.7 is derived from Hinman (1978), and several applications are given in the Appendix, that can be read after Chapter 7. For more information about critical ordinals, see Levy (1979).

Chapter 7

Omega

Up to this point we have been dealing only with finite sets. This situation comes from the fact that at this stage of the system $\mathbb{G}$ there is only one primitive set, the emptyset, and all the operations introduced by the preceding rules produce finite sets when applied to finite set. In this chapter we introduce a new set ω that contains exactly the natural numbers. We discuss first the primitive construction that supports the set ω. As usual, the formal rule introduces the set with an axiom derived from the primitive construction. Studying the natural numbers is not exactly the motivation for this extension, for the predicate **NT** provides sufficient inductive and recursive machinery. The real significance of the set ω comes from the fact that it is an infinite set from which we can reach the universe of the denumerable sets (see also the Appendix).

7.1 The Set Omega

The definition of the predicate **NT** in Chapter 4 was a simple application of the set induction rule. Now we want to introduce a set ω that contains the natural numbers, and this requires a new rule with a supporting primitive construction. This construction is actually independent of the predicate **NT**, although we intend to take advantage of the predicate in the formulation of the formal omega axiom.

We start the construction with the infinite sequence $\emptyset, S(\emptyset), S(S(\emptyset)), \ldots$ which we call the *fundamental sequence.* This is, of course, an informal construction that determines an infinite nonterminating sequence. The sequence is generated by the following rule: If X has been generated, then generate $S(X)$.

We take as ω the *completion* of the fundamental sequence, which means that the elements of ω are exactly all objects that occur in the fundamental sequence. This is an objective construction independent of the universe. It was controversial at some stage in the history of mathematics, but by now is a standard mathematical construction.

Note that the fundamental sequence and its completion are independent of the predicate **NT**. Still, there is a strong relation between the two notions. In fact, it is clear that every element in the fundamental sequence is a natural number. This follows because the rule that generates the fundamental sequence (from X get $S(X)$) preserves the property of being a natural number. Furthermore, every natural number occurs in the fundamental sequence, as can be shown by an informal numerical mathematical induction.

The preceding argument relating the set ω with the predicate **NT** is strictly informal, but we can use it to motivate the omega axiom.

The Omega Rule We introduce a set ω which satisfies the following axiom.

The Omega Axiom:

$$X \in \omega \equiv \mathbf{NT}(X).$$

Remark 7.1 The preceding axiom is not an extensional definition of ω, but an assertion that we assume to be true for any set X in the universe. The validity of the axiom is derived from the primitive construction above, where ω is the completion of the fundamental sequence.

Theorem 7.1 *The set ω is a limit ordinal.*

PROOF. We know that all elements of ω are natural numbers, hence they are ordinals. If $X \in \omega$ and $Y \in X$, then X is a natural number, hence Y is also a natural number, so $Y \in \omega$. This means that ω is transitive, hence is an ordinal. If $\omega = S(X)$, then X is a natural number, hence ω is also a natural number and $\omega \in \omega$, which is a contradiction since ω is an ordinal. □

EXERCISE 7.1 Let α be a limit ordinal. Prove that $\omega \subseteq \alpha$.

In dealing with natural numbers it is convenient to use special variables, as we did before with ordinals. We shall use letters $\mathsf{i}, \mathsf{j}, \mathsf{m}, \mathsf{n}, \mathsf{p}, \mathsf{q}$, etc., as numerical variables. When these variables occur free in a term they are essentially set variables that we intend to restrict to natural numbers. When such variables are bound, the meaning of the terms is affected. We discussed this situation at the time ordinal variables were introduced, and the same interpretation applies here. Of particular importance is the fact that non-local quantification over the natural numbers is admissible in the definition of operations and predicates. The reason is that a non-local expression $(\forall \mathsf{n})$ is equivalent to $(\forall \mathsf{n} \in \omega)$, and $(\exists \mathsf{n})$ is equivalent to $(\exists \mathsf{n} \in \omega)$.

Note that the operations $+_{\mathbf{or}}$ and $+_{\mathbf{nt}}$ are equivalent when applied to natural numbers, and we can use $+_{\mathbf{or}}$ even in cases where the arguments are natural numbers. Similarly for $\times_{\mathbf{or}}$ and $\times_{\mathbf{nt}}$.

EXERCISE 7.2 Let f be a set operation such that $f(\alpha)$ is an ordinal for every ordinal α. Prove that if α is an arbitrary ordinal there is a natural number n such that $f(\alpha +_{\mathbf{or}} \mathsf{n}) \leq_{\mathbf{or}} f(\alpha +_{\mathbf{or}} \mathsf{m})$ for every ordinal m.

EXAMPLE 7.1 From ω we can derive many new limit ordinals. For example, $\omega +_{\mathbf{or}} \omega, \omega +_{\mathbf{or}} \omega +_{\mathbf{or}} \omega, \ldots$ are all limit ordinals. Similarly, $\omega \times_{\mathbf{or}} \omega, \omega \times_{\mathbf{or}} \omega \times_{\mathbf{or}} \omega \ldots$ are all limit ordinals. Later, we show how normal operations can be used to generate limit ordinals of great complexity. □

Once the set ω is available the rule of numerical primitive recursion takes a new dimension, for now we can collect the values of any iteration at level ω. Hence, if the values $f(0), f(1), ..., f(\mathsf{n}), ...$ are available via some set operation f we can introduce $(\lambda\mathsf{n} \in \omega)f(\mathsf{n})$, which is a function, and also $\{f(\mathsf{n}) : \mathsf{n} \in \omega :\} = f[\omega]$, which is the range of f restricted to ω .

7.1.1 $\mathbf{fpw}(Y) = \bigcup\{\mathbf{fs}(\mathsf{n}, Y) : \mathsf{n} \in \omega :\}$ *(the set of all finite subsets of Y).*

Theorem 7.2 *If X and Y are arbitrary sets, then the following conditions are equivalent:*

(i) $X \in \mathbf{fpw}(Y)$.

(ii) $\mathbf{FI}(X) \wedge X \subseteq Y$.

PROOF. Immediate from Theorem 5.9. □

Corollary 7.2.1 *If Y be a finite set, then:*

(i) $\mathbf{fpw}(Y) = \mathbf{tc}_{\mathbf{FI}}(Y) \cup \{Y\}$.

(ii) $\mathbf{fpw}(Y)$ *is finite.*

PROOF. Part (i) follows from Theorems 7.2 and 4.6. Part (ii) follows from Corollary 4.3.2. □

Theorem 7.3 *Let Y be a set such that* $Y \subseteq \mathbf{fpw}(Y)$. *If* $X \in Y$ *and X is well-founded, then X is hereditarily finite.*

PROOF. With Y fixed let $p(X) \equiv X \in Y \rightarrow \mathbf{HF}(X)$. We prove by **WF**-induction that $p(X)$ holds for every well-founded set X. In order to prove $p(X)$ we assume $X \in Y$, and from the assumption on Y this means that X is finite and $X \subseteq Y$. Hence, if $V \in X$, then $V \in Y$ and $p(V)$ holds by the induction hypothesis, so V is hereditarily finite. We conclude that X is hereditarily finite. □

Corollary 7.3.1 *Let Y be a well-founded set such that* $\mathbf{fpw}(Y) = Y$. *If X is an arbitrary set, then the following conditions are equivalent:*

(i) $X \in Y \wedge \mathbf{WF}(X)$.

(ii) $\mathbf{HF}(X)$.

PROOF. The implication from (i) to (ii) is immediate from Theorem 7.3. To prove the implication from (ii) to (i) we note that $\mathbf{HF}(X)$ implies $\mathbf{WF}(X)$, and use **HF**-induction on X to prove that $X \in Y$. We assume that X is hereditarily finite, hence if $V \in X$, then V is finite and $V \in Y$ by the induction hypothesis. This means $X \subseteq Y$, hence $X \in \mathbf{fpw}(Y) = Y$. □

Corollary 7.3.2 *If X, Y are sets where $\mathbf{fpw}(Y) = Y$ and Y is well-founded, then the following conditions are equivalent:*

(i) $X \in Y$.

(ii) $\mathbf{HF}(X)$.

PROOF. The implication from (i) to (ii) follows from Corollary 7.3.1, noting that since $X \in Y$, then X is well-founded. The implication from (ii) to (i) is clear from Corollary 7.3.1. □

Theorem 7.4 *The set ω is a critical ordinal.*

PROOF. From the definition of g in Section 6.7 it is clear that $g(\mathsf{n})$ is a natural number for every natural number n. It follows that $\ll \mathsf{n}, \mathsf{m} \gg$ is a natural number for natural numbers n and m. □

Theorem 7.5 *If n is a natural number, then $\mathsf{n} +_{\mathbf{or}} \omega = \omega$.*

PROOF. $\mathsf{n} +_{\mathbf{or}} \omega = \mathbf{sup}(\{\mathsf{n} +_{\mathbf{or}} \mathsf{m} : \mathsf{m} <_{\mathbf{or}} \omega :\}) = \omega$. □

Corollary 7.5.1 *If κ is an arbitrary ordinal, then $\kappa +_{\mathbf{or}} (\kappa \times_{\mathbf{or}} \omega) = \kappa \times_{\mathbf{or}} \omega$.*

PROOF. Noting that $\kappa \times_{\mathbf{or}} 1 = \kappa$, we have $\kappa +_{\mathbf{or}} (\kappa \times_{\mathbf{or}} \omega) = (\kappa \times_{\mathbf{or}} 1) +_{\mathbf{or}} (\kappa \times_{\mathbf{or}} \omega) = \kappa \times_{\mathbf{or}} (1 +_{\mathbf{or}} \omega) = \kappa \times_{\mathbf{or}} \omega$. □

EXERCISE 7.3 Let α be an infinite ordinal. Prove:

(a) If n is a natural number, then $\mathsf{n} +_{\mathbf{or}} \alpha = \alpha$.

(b) If κ is an arbitrary ordinal, then $\kappa +_{\mathbf{or}} (\kappa \times_{\mathbf{or}} \alpha) = \kappa \times_{\mathbf{or}} \alpha$.

EXAMPLE 7.2 In this example we prove a simple version of the well-known Cantor-Bernstein theorem. Consider a function F where $\mathbf{do}(F) = A$, and $\mathbf{ra}(F) \subseteq B \subseteq A$. We assume that F is 1–1 in the sense of Definition 2.3.11. We want to construct a function F' where $\mathbf{do}(F') = A$, $\mathbf{ra}(F') = B$, and F' is also 1–1. We use primitive recursion to introduce the sequence of sets $A_0, A_1, ..., A_{\mathsf{n}}, ...$ such that

$$A_0 = B \setminus \mathbf{ra}(F) = B \setminus F[A],$$

$$A_{\mathsf{n}+1} = F[A_{\mathsf{n}}].$$

Now we use ω to introduce a set A' and a function F' such that:

$$A' = \bigcup\{A_{\mathsf{n}} : \mathsf{n} \in \omega :\}$$

$$F' = \{<V, W> : V \in A \wedge W \in B : ((V \in A' \wedge V = W) \vee (V \notin A' \wedge \mathbf{ap}(F, V) = W))\}.$$

Clearly, F' is a function where $\mathbf{do}(F') = A$ and $\mathbf{ra}(F') \subseteq B$. Actually, $\mathbf{ra}(F') = B$, for if $W \in B$ and $W \notin A'$, then $W \in \mathbf{ra}(F)$ (otherwise, $W \in A_0$), hence $\mathbf{ap}(F, V) = W$, where $V \notin A'$. Finally, we can prove that F' is 1–1. Assume $<V, W> \in F'$ and $<V', W> \in F'$ where $V \neq V'$. We may assume $V \in A'$, $V = W$, $V' \notin A'$, and $\mathbf{ap}(F, V') = W$. From $V' \notin A'$ we conclude $\mathbf{ap}(F, V') \notin A'$ (since F is 1–1) and this contradicts $W \in A'$. □

EXERCISE 7.4 Let F, A, B and A' be as in Example 7.2. Prove:

(a) $A' \subseteq B$.

(b) If $W \in B \setminus A'$, then there is $V \in A \setminus A'$ such that $\mathbf{ap}(F, V) = W$.

(c) If $V \in A$ and $\mathbf{ap}(F, V) \in A'$, then $V \in A'$.

EXAMPLE 7.3 The following is an application of the construction in Example 7.2. Let Z be a non-empty set and G a 1–1 function where $\mathbf{do}(G) = Z \times Z$, and $\mathbf{ra}(G) \subseteq Z$. We want to construct a 1–1 function G', where $\mathbf{do}(G') = Z \times Z$ and $\mathbf{ra}(G') = Z$. In order to apply the construction in Example 7.2 we fix $Y \in Z$, and set $A = Z \times Z$, $B = Z \times \{Y\} \subseteq A$. First, we change G to F, where $\mathbf{do}(F) = A$, and $\mathbf{ap}(F, X) = <\mathbf{ap}(G, X), Y>$ whenever $X \in A$, hence F is 1–1 and $\mathbf{ra}(F) \subseteq B$. By the construction in Example 7.2, there is a 1–1 function F' where $\mathbf{do}(F') = A$ and $\mathbf{ra}(F') = B$. Finally, we set a 1–1 function G' where $\mathbf{do}(G') = A$ and $\mathbf{ap}(G', X) = [\mathbf{ap}(F', X)]_1$ whenever $X \in Z \times Z$. Clearly, $\mathbf{ra}(G') = Z$. □

Theorem 7.6 *If κ is a critical ordinal, then there is a 1–1 function F such that* $\mathbf{do}(F) = \kappa$ *and* $\mathbf{ra}(F) = \kappa \times \kappa$.

PROOF. From Example 6.8 we get a 1–1 function G such that $\mathbf{do}(G) = \kappa \times \kappa$ and $\mathbf{ra}(F) \subseteq \kappa$. Using the construction of Example 7.3 and we get a 1–1 function G' such that $\mathbf{do}(G') = \mathbf{do}(G) = \kappa \times \kappa$ and $\mathbf{ra}(G') = \kappa$. Finally, we take as F the inversion of G', namely:

$$F = \{<Y, X> : Y \in \mathbf{ra}(G') \wedge X \in \mathbf{do}(G') : <X, Y> \in G.\}$$

Clearly, F is 1–1, $\mathbf{do}(F) = \mathbf{ra}(G') = \kappa$, and $\mathbf{ra}(F) = \mathbf{do}(G') = \kappa \times \kappa$. □

7.2 Local Induction and Normal Operations

From the omega axiom and the recursive techniques derived from **NT** we can develop a substantial part of classical mathematics. We do not intend to pursue this approach, since this type of translation is well covered in the literature. We prefer to move to applications that are set theoretical in nature. In many cases the crucial construction consists of generating a sequence of sets $X_0, X_1, ..., X_n..., X_\omega, ..., X_\alpha, ...$ and taking the limit via the union of the elements in the sequence.

An important application of this type of process is the derivation of local induction. Rather than an inductive predicate, as is the case with the set induction rule of Chapter 3, here we are concerned with a set that satisfies similar inductive properties. A very general approach to this construction is to start with a given set operation g and determine in which conditions there is a set X such that $g(X) = X$. If X exists and it is minimal, then it has very useful inductive properties. The existence of such a set X requires a number of assumptions about the operation g.

Let g be a unary set operation. If X is a set such that $g(X) \subseteq X$ we say that g is *closed at* X. If $X \subseteq g(X)$ we say that g is *reflexive* at X. If g is both closed and reflexive at X, it follows that $g(X) = X$, and we say that X is a *fixed point* of g.

EXAMPLE 7.4 Consider the operation $g(X) = \bigcup X$. If the set X is transitive, then g is closed at X. If α is a limit ordinal, then α is a fixed point of g. □

EXAMPLE 7.5 Consider the operation $g(X) = S(X)$. Clearly, g is reflexive at any set X. If X is a well-founded set, then g is not closed at X. □

A set operation g is *monotonic* if whenever X and Y are sets such that $X \subseteq Y$, then $g(X) \subseteq g(Y)$. The operation g in Example 7.4 is monotonic. The operation g is Example 7.5 is not monotonic. If $g(X) \subset g(Y)$ whenever $X \subset Y$, we say that g is *strictly monotonic.* Note that monotonicity is not a predicate in the system $\mathbb{G}$, even if g is a fixed basic operation, for it involves quantification over the universe. We consider this to be a *condition* that can be verified in specific cases (see the reduction theorems in Chapter 10).

EXAMPLE 7.6 The operation $g(X) = X \cup \bigcup X$ is monotonic. A set X is a fixed point of g if and only if X is transitive. □

Remark 7.2 The problem in this section is to determine in which conditions a monotonic operation g has a fixed point. We approach this problem below assuming that g is monotonic and satisfies continuity conditions (see Remark 7.3). In Chapter 8 we take a different approach and show that if g is closed at a set X, then there is a minimal fixed point Z such that $Z \subseteq X$.

EXAMPLE 7.7 The operation **fpw** in 7.1.1 is strictly monotonic, for if $X \subset Y$ and $V \in Y$, $V \notin X$, then $\{V\} \in \mathbf{fpw}(Y)$, but $\{V\} \notin \mathbf{fpw}(X)$. If **fpw** is closed at X, this means that whenever $Y \subseteq X$ and Y is finite, then $Y \in X$. Hence, if X is a finite well-founded set, then **fpw** is not closed at X. If **fpw** is reflexive at X, and X is well-founded, then all the elements of X are well-founded sets (Theorem 7.3). If **fpw** is closed at X, then $Y \in X$ whenever Y is finite and $Y \subseteq X$. Hence, if X is finite and well-founded, then **fpw** is not closed at X. From Corollary 7.3.2 it follows that whenever X is a fixed point of **fpw** and X is well-founded, then X is the set of all hereditarily finite sets. □

We associate with g a binary set operation $g^{\odot}$ defined by primitive ordinal recursion as follows:

7.2.1-1 $g^{\odot}(0, Y) = Y$.

7.2.1-2 $g^{\odot}(S(\alpha), Y) = g(g^{\odot}(\alpha, Y))$.

7.2.1-3 $g^{\odot}(\alpha, Y) = \bigcup\{g^{\odot}(\gamma, Y) : \gamma <_{\mathbf{or}} \alpha :\}$ (α a limit ordinal).

We say that $g^{\odot}$ is the *iteration operation from g*.

EXERCISE 7.5 Prove that if Y is a fixed point of g, then $g^{\odot}(\alpha, Y) = Y$ for every ordinal α.

Theorem 7.7 *If If g is monotonic, α is an ordinal, and Y, X are sets such that $Y \subseteq X$ and g is closed at X, then $g^{\odot}(\alpha, Y) \subseteq X$.*

PROOF. The proof is by ordinal induction on α, with X, Y fixed. The case $\alpha = 0$ follows from the assumption $Y \subseteq X$. If $g^{\odot}(\alpha, Y) \subseteq X$, then $g^{\odot}(S(\alpha), Y) = g(g^{\odot}(\alpha, Y)) \subseteq g(X) \subseteq X$. If α is a limit ordinal and $\gamma <_{\mathbf{or}} \alpha$, then $g^{\odot}(\gamma, Y) \subseteq X$ by the induction hypothesis. From definition 7.2.1-3 it follows that $g^{\odot}(\alpha, Y) \subseteq X$. □

Theorem 7.8 *Assume the operation g is monotonic, α is an ordinal, and g is reflexive at the set Y. Then:*

(i) *g is reflexive at $g^{\odot}(\alpha, Y)$.*

(ii) *If $\gamma <_{\mathbf{or}} \alpha$, then $g^{\odot}(\gamma, Y) \subseteq g^{\odot}(\alpha, Y)$.*

(iii) *$Y \subseteq g^{\odot}(\alpha, Y)$.*

PROOF. The proof of (i) is by ordinal induction on α. The case $\alpha = 0$ follows from the assumption on Y. If the relation holds for α, we compute using monotonicity:

$$g^{\odot}(S(\alpha), Y) = g(g^{\odot}(\alpha, Y)) \subseteq g(g(g^{\odot}(\alpha, Y))) = g(g^{\odot}(S(\alpha), Y)).$$

If α is a limit ordinal and $\gamma <_{\mathbf{or}} \alpha$ we have by the induction hypothesis and monotonicity,

$$g^{\odot}(\gamma, Y) \subseteq g(g^{\odot}(\gamma, Y)) \subseteq g(\bigcup\{g^{\odot}(\gamma, Y) : \gamma <_{\mathbf{or}} \alpha :\}) = g(g^{\odot}(\alpha, Y)),$$

hence $g^{\odot}(\alpha, Y) \subseteq g(g^{\odot}(\alpha, Y))$. Part (ii) follows again by ordinal induction on α. The case $\alpha = 0$ is trivial. If $\gamma <_{\mathbf{or}} S(\alpha)$ we have $g^{\odot}(\gamma, Y) \subseteq g^{\odot}(\alpha, Y)$ by the induction hypothesis, and $g^{\odot}(\alpha, Y) \subseteq g^{\odot}(S(\alpha), Y)$ by part (i). Finally, if α is a limit ordinal and $\gamma <_{\mathbf{or}} \alpha$, we have $g^{\odot}(\gamma) \subseteq g^{\odot}(\alpha, Y)$ from definition 7.2.1-3. Part (iii) follows from (ii), for $Y = g^{\odot}(0, Y)$. □

EXERCISE 7.6 Assume g is monotonic, α is a limit ordinal and g is reflexive at Y. Prove that $g^{\odot}(\alpha, Y) = \bigcup\{g^{\odot}(S(\gamma), Y) : \gamma <_{\mathbf{or}} \alpha :\}$.

Theorem 7.9 *Assume g is a monotonic operation, g is reflexive at Y, and κ is an ordinal such that $g^{\odot}(\kappa, Y)$ is a fixed point of g. Then:*

(i) *If $\kappa <_{\mathbf{or}} \alpha$, then $g^{\odot}(\kappa, Y) = g^{\odot}(\alpha, Y)$.*

(ii) *If X is a fixed point of g and $Y \subseteq X$, then $g^{\odot}(\kappa, Y) \subseteq X$.*

Part (i) follows by ordinal induction on α. The case $\alpha = 0$ is trivial. If $\kappa <_{\mathbf{or}} S(\alpha)$, then by the induction hypothesis we have $g^{\odot}(\kappa, Y) = g^{\odot}(\alpha, Y)$, hence $g^{\odot}(\kappa, Y) = g(g^{\odot}(\alpha, Y)) = g^{\odot}(S(\alpha), Y)$. If α is a limit ordinal and $\kappa \leq_{\mathbf{or}} \gamma <_{\mathbf{or}} \alpha$, we have $g^{\odot}(\kappa, Y) = g^{\odot}(\gamma, Y)$. From Definition 7.2.1-3 it follows that $g^{\odot}(\alpha, Y) = g^{\odot}(\kappa, Y)$. Part (ii) follows from Theorem 7.7. □

Theorem 7.10 *Assume the operation g is monotonic, κ is an ordinal, g is reflexive at Y, and g is closed at $g^{\odot}(\kappa, Y)$. Then $g^{\odot}(\kappa, Y)$ is a fixed point of g.*

PROOF. Immediate from Theorem 7.8 (i).

Remark 7.3 When κ is a limit ordinal, the condition in Theorem 7.10 means that $g(g^{\odot}(\kappa, Y)) = \bigcup\{g^{\odot}(\gamma, Y) : \gamma <_{\mathbf{or}} \kappa :\} = \bigcup\{g^{\odot}(S(\gamma), Y) : \gamma <_{\mathbf{or}} \kappa :\} = \bigcup\{g(g^{\odot}(\gamma, Y)) : \gamma <_{\mathbf{or}} \kappa :\}$, and this is a continuity property of the operation g.

EXAMPLE 7.8 Consider the operation $g(X) = X \cup \bigcup X$ of Example 7.6 and assume κ is a limit ordinal. Note that g is reflexive at any set Y. The closure condition in Theorem 7.10 takes the form, $X \cup \bigcup\bigcup X \subseteq \bigcup X$, where $X = \{g^{\odot}(\gamma, Y) : \gamma <_{\mathbf{or}} \kappa :\}$, so if V is a set such that $V \in X$, then $V \cup \bigcup V \in X$. To show that the closure condition holds, note that if $Z \in W \in V \in X$, then $Z \in \bigcup V$, so $Z \in \bigcup X$. We conclude that $g^{\odot}(\kappa, Y)$ is a fixed point of g. In particular, $g^{\odot}(\omega, Y)$ is a fixed point of g. □

EXERCISE 7.7 Let g be the operation in Example 7.8 and Y an arbitrary set. Prove:

(a) $Y \subseteq g^{\odot}(\omega, Y)$.

(b) $g^{\odot}(\omega, Y)$ is transitive.

(c) If $Y \subseteq X$ and X is transitive, then $g^{\odot}(\omega, Y) \subseteq X$.

(d) If Y is well-founded, then $g^{\odot}(\omega, Y) = \mathbf{tc}(Y)$.

In some cases it is convenient to apply Theorem 7.10 using the notation of nets (Definition 2.3.12). We recall that if F is a net, then $\mathbf{do}(F)$ is a directed set. If $\mathbf{do}(F)$ is an ordinal, then we say that F is an *ordinal net.*

If g is a monotonic operation, g is reflexive at Y, and κ is an ordinal, we can introduce the ordinal net F where $\mathbf{do}(F) = \kappa$, and $\mathbf{ap}(F, \gamma) = g^{\odot}(\gamma, Y)$ holds for every $\gamma <_{\mathbf{or}} \kappa$. We say that F is the κ-net *induced by g at Y*. Note that the relation in Theorem 7.8 (ii) is essential for this construction. Here we are interested in ordinal nets where the domain is a limit ordinal.

Now the condition in Theorem 7.10 in the sense that g is closed at $g^{\odot}(\kappa, Y)$ can be expressed in the equivalent form, $g(\bigcup \mathbf{ra}(F)) \subseteq \bigcup \mathbf{ra}(F)$, where κ is a limit ordinal and F is the κ-net induced by g at Y. Note that this translation is valid under the assumptions above, namely, that g is monotonic, $Y \subseteq g(Y)$, and κ is a limit ordinal.

Corollary 7.10.1 *Assume the operation g is monotonic, κ is a limit ordinal, g is reflexive at Y, and $g(\bigcup \mathbf{ra}(F)) \subseteq \bigcup \mathbf{ra}(F)$, where F is the κ-net induced by g at Y. Then, $g^{\odot}(\kappa, Y)$ is a fixed point of g.*

PROOF. Immediate from Theorem 7.10, noting that $g(\bigcup \mathbf{ra}(F)) \subseteq \bigcup \mathbf{ra}(F)$ implies that g is closed at $g^{\odot}(\kappa, Y)$. □

EXAMPLE 7.9 We can show now that $\mathbf{fpw}^{\odot}(\omega, \emptyset)$ is a fixed point of $\mathbf{fpw}$. If F is the ω-net induced by $\mathbf{fpw}$ at $\emptyset$, we must show that $\mathbf{fpw}(\bigcup \mathbf{ra}(F)) \subseteq \bigcup \mathbf{ra}(F)$. If $Z \in \mathbf{fpw}(\bigcup \mathbf{ra}(F))$, then Z is finite and $Z \subseteq \bigcup \mathbf{ra}(F)$. From Theorem 4.2 it follows that there is n such that $Z \subseteq \mathbf{fpw}^{\odot}(\mathsf{n}, \emptyset)$, hence $Z \in \mathbf{fpw}^{\odot}(S(\mathsf{n}), \emptyset)$. It follows that $Z \in \bigcup \mathbf{ra}(F)$. We conclude that $\mathbf{fpw}^{\odot}(\omega, \emptyset)$ is a fixed point of **fpw**. Since $\mathbf{fpw}^{\odot}(\omega, \emptyset)$ is clearly well-founded, from Corollary 7.3.2 it follows that $\mathbf{fpw}^{\odot}(\omega, \emptyset)$ is the set of all hereditarily finite sets. □

EXERCISE 7.8 Prove that $\mathbf{fpw}^{\odot}(\mathsf{n}, \emptyset)$ is a finite set for every natural number n.

Now we change our assumptions about g as follows. If $g(\alpha)$ is an ordinal for every ordinal α, we say that g is an *ordinal operation.* If $g(\gamma) \leq_{\mathbf{or}} g(\alpha)$, whenever $\gamma \leq_{\mathbf{or}} \alpha$ we say that g is *ordinal monotonic.* If $g(\gamma) <_{\mathbf{or}} g(\alpha)$ whenever $\gamma <_{\mathbf{or}} \alpha$, we say that g is *strictly ordinal monotonic.* The definition of closed, reflexive, and fixed point are the same as before, with the understanding that now we are interested only in ordinals.

EXAMPLE 7.10 The operation in Example 7.4 is ordinal monotonic, but it is not strictly monotonic, for if α is a limit ordinal, we have $g(\alpha) = g(S(\alpha)) = \alpha$. Clearly, the operation g in Example 7.5 is strictly monotonic. The operation **fpw** in Example 7.7 is not an ordinal operation. □

EXAMPLE 7.11 The ordinal operation $g(\alpha) = \omega +_{\mathbf{or}} \alpha$ is strictly monotonic. The ordinal $\omega \times_{\mathbf{or}} \omega$ is a fixed point of g (Corollary 7.5.1). If $\omega \times_{\mathbf{or}} \omega <_{\mathbf{or}} \gamma$, then γ is also a fixed point of g. □

EXERCISE 7.9 Assume the ordinal operation g is strictly monotonic. Prove that g is reflexive at every ordinal α.

The definition of the set operation $g^{\odot}$ remains the same, for it is independent of any particular assumption about g. Still, now we are interested in the application of $g^{\odot}$ to ordinals, say in the form $g^{\odot}(\alpha, \beta)$. Since the definition of $g^{\odot}$ involves only the operations successor and union, it is clear that $g^{\odot}(\alpha, \beta)$ is an ordinal.

It should be clear now that the only change here is that we consider a partial universe consisting of only the ordinals, and disregard the global universe. By doing this we extend the range of application of the theory. For example, the operation $g(X) = S(X)$ is not monotonic over the universe, and it is excluded from Theorem 7.10. On the other hand, g is strictly monotonic as an ordinal operation, for in case $\gamma <_{\mathbf{or}} \alpha$, then $S(\gamma) <_{\mathbf{or}} S(\alpha)$. This means that Theorem 7.10 applies to g when Y is an ordinal. On the other hand, note that there is no ordinal β such that g is closed at β.

Finally, in dealing with ordinals we write $\leq_{\mathbf{or}}$ rather than $\subseteq$, and $<_{\mathbf{or}}$ is equivalent to $\in$. In the same vain, we use $\mathbf{sup}(X)$ rather than $\bigcup X$ when X is a set of ordinals.

From these considerations it follows that Theorem 7.7 still holds if we assume g is a monotonic ordinal operation, and X, Y are ordinals. Similarly, Theorems 7.8, 7.9 and 7.10 hold under the same assumptions and furthermore $Y \leq_{\mathbf{or}} g(Y)$.

As before, Theorem 7.10 provides a general method to prove that $g^{\odot}(\kappa, \beta)$ is the fixed point of the ordinal operation g. The relation in Example 7.10 can be replaced with an equivalent condition in terms of nets, as explained in Example 7.9. In fact, it is convenient to use this formulation to make more explicit that the relation is a form of continuity (Remarks 7.2 and 7.3).

If κ is a limit ordinal, we say that the ordinal operation g is κ-continuous if whenever F is an ordinal net where $\mathbf{do}(F) = \kappa$ and $g(\mathbf{ap}(F, \gamma)) \leq_{\mathbf{or}} \mathbf{ap}(F, S(\gamma))$ holds for every $\gamma <_{\mathbf{or}} \kappa$, then $g(\mathbf{sup}(\mathbf{ra}(F))) \leq_{\mathbf{or}} \mathbf{sup}(\mathbf{ra}(F))$. In applications we are interested in those κ-nets induced by the operation g at some ordinal β such that $\beta \leq_{\mathbf{or}} g(\beta)$.

EXAMPLE 7.12 For a given limit ordinal κ, consider the net F such that $\mathbf{do}(F) = \kappa$ and $\mathbf{ap}(F, \gamma) = \gamma$ for every $\gamma <_{\mathbf{or}} \kappa$. It follows that $S(\mathbf{ap}(F, \gamma)) = \mathbf{ap}(F, S(\gamma))$. On the other hand $S(\mathbf{sup}(\mathbf{ra}(F))) = S(\kappa) \not\leq_{\mathbf{or}} \kappa = \mathbf{sup}(\mathbf{ra}(F))$, so the operation S is not κ-continuous. □

Theorem 7.11 *Assume κ is a limit ordinal and the ordinal operation g is ordinal monotonic, κ-continuous, and $\beta \leq_{\mathbf{or}} g(\beta)$. Then:*

(i) *$g^{\odot}(\kappa, \beta)$ is a fixed point of g.*

(ii) *If $\kappa <_{\mathbf{or}} \alpha$, then $g^{\odot}(\kappa, \beta) = g^{\odot}(\alpha, \beta)$.*

PROOF. To prove (i), let F be the ω-net induced by g at β. By κ-continuity we have $g(\mathbf{sup}(\mathbf{ra}(F))) \leq_{\mathbf{or}} \mathbf{sup}(\mathbf{ra}(F))$, and by Corollary 7.10.1 this means that $g^{\odot}(\kappa, \beta)$ is a fixed point of g. Part (ii) follows from Theorem 7.9 (ii).□

Theorem 7.12 *Assume the ordinal operation g is ordinal monotonic, and furthermore, satisfies the following condition whenever α is a limit ordinal: $g(\alpha) = \mathbf{sup}(\{g(\gamma') : \gamma' <_{\mathbf{or}} \alpha :\})$. Then, g is κ-continuous for every limit ordinal κ.*

PROOF. Let F be an ordinal net where $\mathbf{do}(F) = \kappa$ and $g(\mathbf{ap}(F, \gamma)) \leq_{\mathbf{or}} \mathbf{ap}(F, S(\gamma))$ holds for every $\gamma <_{\mathbf{or}} \kappa$. We set $\alpha = \mathbf{sup}(\mathbf{ra}(F))$ to show that $g(\alpha) \leq_{\mathbf{or}} \alpha$. We consider two cases. If $\alpha \in \mathbf{ra}(F)$, then $\alpha = \mathbf{ap}(F, \gamma)$ for some $\gamma <_{\mathbf{or}} \kappa$, hence $g(\alpha) = g(\mathbf{ap}(F, \gamma)) \leq_{\mathbf{or}} \mathbf{ap}(F, S(\gamma)) = \alpha$. In the second case we assume $\alpha \notin \mathbf{ra}(F)$, hence α is a limit ordinal, and $g(\alpha) = \mathbf{sup}(\{g(\gamma') : \gamma' <_{\mathbf{or}} \alpha :\})$. If $\gamma' <_{\mathbf{or}} \alpha$, then $\gamma' <_{\mathbf{or}} \mathbf{ap}(F, \gamma)$ for some $\gamma <_{\mathbf{or}} \kappa$, hence $g(\gamma') \leq_{\mathbf{or}} g(\mathbf{ap}(F, \gamma)) \leq_{\mathbf{or}} \mathbf{ap}(F, S(\gamma)) <_{\mathbf{or}} \alpha$. It follows that $g(\alpha) \leq_{\mathbf{or}} \alpha$. □

EXAMPLE 7.13 Let κ be a fixed ordinal and define $g(\alpha) = \kappa +_{\mathbf{or}} \alpha$. Clearly, g satisfies the condition in Theorem 7.12, so g is ω-continuous. By Theorem 7.11 it follows that $g^{\odot}(\omega, 0)$ is the minimal fixed point of g. Noting that $g^{\odot}(0,0) = 0$, $g^{\odot}(1,0) = \kappa$, $g^{\odot}(2,\kappa) = \kappa +_{\mathbf{or}} \kappa$, $g^{\odot}(3,\kappa) = \kappa +_{\mathbf{or}} \kappa +_{\mathbf{or}} \kappa$, etc., we conclude that $g^{\odot}(\omega, 0) = \mathbf{sup}(\{\kappa \times_{\mathbf{or}} \mathsf{n} : \mathsf{n} <_{\mathbf{or}} \omega :\}) = \kappa \times_{\mathbf{or}} \omega$. □

EXAMPLE 7.14 The ω-continuity property provides a convenient approach to prove that $g^{\odot}(\omega, \beta)$ is a fixed point of a given ordinal monotonic operation g. Still, it depends on the condition $\beta \leq_{\mathbf{or}} g(\beta)$, which may fail in some cases. For example, if $g(\alpha) = 0$ for every α, then g is ordinal monotonic, but $1 \leq_{\mathbf{or}} g(1)$ fails. This problem can be solved if we assume strict monotonicity rather than simple monotonicity, for in this case we have $g(\beta) <_{\mathbf{or}} g(S(\beta))$, hence $S(g(\beta)) \leq_{\mathbf{or}} g(S(\beta))$, and using ordinal induction on β we can prove that $\beta \leq_{\mathbf{or}} g(\beta)$ for every ordinal β. □

The preceding example suggests the following definition. We say that the ordinal operation g is a *normal operation* if g is strictly ordinal monotonic and ω-continuous.

If g is a normal operation, then $g^{\odot}(\omega, 0)$ is the least fixed point of g. Noting that $S(g^{\odot}(\omega, 0)) \leq_{\mathbf{or}} g(S(g^{\odot}(\omega, 0)))$ it follows that $g^{\odot}(\omega, S(g^{\odot}(\omega, 0)))$ is the least fixed point of g greater than $g^{\odot}(\omega, 0)$. In general, if β is a fixed point of g, then $g^{\odot}(\omega, S(\beta))$ is the least fixed point of g greater than β.

It is clear that the preceding construction can be used to generate systematically fixed points of a given normal operation g. We formalize this in the form of another operation G defined by primitive ordinal recursion as follows:

$$G(0) = g^{\odot}(\omega, 0).$$

$$G(S(\alpha)) = g^{\odot}(\omega, S(G(\alpha))).$$

$$G(\alpha) = \mathbf{sup}(\{G(\gamma) : \gamma <_{\mathbf{or}} \alpha :\}) \ (\alpha \text{ a limit ordinal}).$$

Theorem 7.13 *Assume the operation g is a normal operation. Then:*

(i) *The operation G is a normal operation.*

(ii) *If α is an ordinal, then $G(\alpha)$ is a fixed point of g.*

PROOF. To prove (i) we need only to show that G is strictly ordinal monotonic. We use ordinal induction on α with γ fixed to show that whenever $\gamma <_{\mathbf{or}} \alpha$, then $G(\gamma) <_{\mathbf{or}} G(\alpha)$. The case $\alpha = 0$ is trivial. We know that $G(S(\alpha))$ is the least fixed point of g greater that $G(\alpha)$, so $G(\alpha) <_{\mathbf{or}} G(S(\alpha))$, and by the induction hypothesis we have $G(\gamma) \leq_{\mathbf{or}} G(S(\alpha))$. The case α is a limit ordinal is clear from the definition of G, noting again that $G(\gamma) <_{\mathbf{or}} G(S(\gamma)) \leq_{\mathbf{or}} G(\alpha)$. The proof of (ii) is also by ordinal induction on α. We need to consider only the case where α is a limit ordinal and prove $g(G(\alpha)) \leq_{\mathbf{or}} G(\alpha)$. This follows by α-continuity of g, noting that if $\gamma <_{\mathbf{or}} \alpha$, then $g(G(\gamma)) = G(\gamma) <_{\mathbf{or}} g^{\odot}(\omega, S(G(\gamma))) = G(S(\gamma))$. □

EXERCISE 7.10 Prove that if δ is a fixed point of the normal operation g, there is $\alpha \leq_{\mathbf{or}} \delta$ such that $G(\alpha) = \delta$.

EXERCISE 7.11 Let g be the ordinal operation in Example 7.11. Prove that $G(\alpha) = (\omega \times_{\mathbf{or}} \omega) +_{\mathbf{or}} \alpha$ for every ordinal α.

7.3 Hereditarily Finite Sets

The predicate **HF** of hereditarily finite sets was introduced in Chapter 4, and a number of results were derived in Chapter 5. In Section 7.2 (Example 7.9) we proved that the set $\mathbf{fpw}^{\odot}(\omega, \emptyset)$ is the set of all hereditarily finite sets, so we define formally the set:

7.3.1 $\mathbf{hf} = \mathbf{fpw}^{\odot}(\omega, \emptyset)$ (*the set of all hereditarily finite sets*).

Theorem 7.14 $\mathbf{fpw}^{\odot}(\mathsf{n}, \emptyset) \in \mathbf{hf}$ *for every natural number* n.

PROOF. From Corollary 7.2.1 it follows that $\mathbf{fpw}^{\odot}(\mathsf{n}, \emptyset)$ is finite, hence $\mathbf{fpw}^{\odot}(\mathsf{n}, \emptyset) \in \mathbf{fpw}(\mathbf{fpw}^{\odot}(\mathsf{n}, \emptyset)) = \mathbf{fpw}^{\odot}(S(\mathsf{n}), \emptyset) \subseteq \mathbf{hf})$. □

Theorem 7.15 *Let p be a binary general predicate, and Z a finite set such that* $(\forall Y \in Z)(\exists \mathsf{m})(\exists V \in \mathbf{fpw}^{\odot}(\mathsf{m}, \emptyset))p(Y, V)$. *Then,* $(\exists \mathsf{n})(\forall Y \in Z)(\exists V \in \mathbf{fpw}^{\odot}(\mathsf{n}, \emptyset))p(Y, V)$.

PROOF. The proof is by **FI**-induction on Z. The case $Z = \emptyset$ is trivial. Otherwise, there is $Y \in Z$, and by the induction hypothesis there is n' such that $(\forall Y' \in Z \ominus Y)(\exists V \in \mathbf{fpw}^{\odot}(\mathsf{n}', \emptyset))p(Y', V)$. On the other hand, we know that there is n'' such that $(\exists V \in \mathbf{fpw}^{\odot}(\mathsf{n}'', \emptyset))p(Y, V)$. So, if we take $\mathsf{n} = \mathsf{n}' \cup \mathsf{n}''$ it follows that $(\forall Y \in Z)(\exists V \in \mathbf{fpw}^{\odot}(\mathsf{n}, \emptyset))p(Y, V)$. □

Corollary 7.15.1 *If p is a binary general predicate, then p is local in* **hf**.

PROOF. Immediate from Theorems 7.14 and 7.15. □

Theorem 7.16 *Assume a set operation f is introduced by an application of the replacement rule, where* **hf** *is closed under the given operations in the rule. Then,* **hf** *is also closed under f.*

PROOF. To prove the theorem we consider an example that can be generalized to any arbitrary application of the replacement rule. We assume f is a binary set operation introduced as follows:

$$f(X_1, X_2) = \{Z \in h(X_1, X_2, Y_1, Y_2) : Y_1 \in g_1(X_1, X_2) \wedge Y_2 \in g_2(X_1, X_2) : p(Z, Y_1, Y_2, X_1, X_2)\}.$$

Our assumption is that **hf** is closed under h, g_1, g_2. We introduce a binary set operation g such that $g(X_1, X_2)) = g_1(X_1, X_2) \times g_2(X_1, X_2)$, and from Corollary 4.7.1 it follows that **hf** is also closed under g. If we set $V = \bigcup\{h(X_1, X_2, [Y]_1, [Y]_2) : Y \in g(X_1, X_2) :\}$, it follows that $f(X_1, X_2) \subseteq V$ holds for arbitrary X_1, X_2. If X_1, X_2 are elements of **hf**, they are hereditarily finite, so $g(X_1, X_2)$ is hereditarily finite and by Theorem 4.7 $h(X_1, X_2, [Y]_1, [Y]_2)$ is hereditarily finite whenever $Y \in g(X_1, X_2)$. We conclude that V is the union of a finite set where all the elements are hereditarily finite, so V is hereditarily finite. It follows that $f(X_1, X_2)$ is hereditarily finite. □

Corollary 7.16.1 *The set* **hf** *is a local universe.*

PROOF. The set **hf** is clearly well-founded. It is also transitive, because an element of an hereditarily finite set is also hereditarily finite. Clearly, **hf** is closed under the insertion operation, and by Theorem 7.16 it is closed under the replacement rule. Furthermore, any general predicate is local in **hf** by Corollary 7.15.1. We conclude that **hf** is a local universe. □

Remark 7.4 Note that Theorems 7.15 and 7.16 are not theorems in the system $\mathbb{G}$. Rather, they describe an infinite number of formal theorems, where the predicate p and the operation f are replaced by concrete basic predicates and operations. In fact, the property of being a local universe cannot be expressed in the language of the theory. □

7.4 Notes

We have used induction to define a number of predicates that play a significant role in the formulation of set theory. Some of them are extremely general, such as the concepts of ordinal number and well-foundedness, and cannot be reduced to the extension of a set. On the other hand, the natural number predicate **NT** is much more restricted, and can be reduced to the extension of the set ω. The fact is that there is an objective process that generates the natural numbers, but there are no similar processes to generate the well-founded sets or the ordinals. Still, the properties of being a well-founded set or an ordinal are objectively defined and can be used with confidence in every application allowed by the rules of the system.

We use a two-step procedure where first we introduce the predicate **NT** and then the set ω. We could have reduced the process to just one rule for ω, at the price of including new axioms of ω-induction and ω-recursion. This would be a distortion of the system, since the original rules of set induction and set recursion provide all the machinery we need.

In standard set theory (say, Zermelo-Fraenkel) the set ω is usually introduced as the intersection of all subsets (of an axiomatically given set) that contain 0 and are closed under the successor operation. This is a very useful axiomatic device, but it is not a definition or construction of ω. The axiom in our omega rule is also an axiomatic device, and it is not intended as a definition or construction of ω. Essentially, we take the natural numbers from the ordinary mathematical practice with the understanding that the particular representation in terms of sets is, to some extent, arbitrary and can be replaced by different conventions, for example, the one described in Example 3.6. In this approach the set ω is the limit of the fundamental sequence that generates the natural numbers, and this process is actually independent of the particular representation.

More philosophical discussions about the nature of the natural numbers are available in the literature. In particular, Hallett (1984) considers in great detail the quasi-theological views of G. Cantor, the creator of set theory.

The application of ω to normal operations comes from Veblen (1908). For a modern presentation, see Levy (1979).

The proof of the general Cantor-Bernstein theorem is given in many places, in particular in Enderton (1977) which contains useful historical information.

Chapter 8

Power-Set and Cardinals

The theory of cardinal numbers is a classical fragment of standard set theory that depends on resources still not available to us. In particular, the power-set construction and the axiom of choice play substantial roles in the theory. In this chapter we introduce a new construction from which these tools can be derived. In this way our system comes very close to standard set theory, although some constructions are still not available.

8.1 Ordinal Permutations

In this section we describe a new permutation rule where a unary set operation Ξ is characterized via formal axioms. The idea is that $\Xi(Z)$ contains all the ordinal permutations of a set Z. We start with a formal definition and proceed to describe the primitive construction from which the formal axioms are derived.

8.1.1 $\mathbf{PM}(Z, F) \equiv \mathbf{FU1}(F) \wedge \mathbf{OR}(\mathbf{do}(F)) \wedge \mathbf{ra}(F) = Z$ (*F is an ordinal permutation of Z*).

In dealing with 1–1 functions we shall use several well-known constructions. For example, if F is a 1–1 function, then the inverse of F is the function $F' = \{<X, Y> : X \in \mathbf{ra}(F) \wedge Y \in \mathbf{do}(F) : <Y, X> \in F\}$. Clearly,

F' is also a 1–1 function. Another useful construction is the composition of given 1–1 functions F_1 and F_2. This is the set

$$F = \{<X,Y> : X \in \mathbf{do}(F_1) \wedge Y \in \mathbf{ra}(F_2) : (\exists Z \in \mathbf{ra}(F_1))<X,Z> \in F_1 \wedge <Z,Y> \in F_2\}.$$

Clearly, F is also a 1–1 function.

We call a *free choice transfinite permutation over a set Z* (**FCTP** over Z), a process where, for every ordinal α, an element $X_\alpha \in Z$ is assigned, the assignment being made by a free choice. We assume the choices are made (almost) simultaneously, with the only restriction that the choice of the element X_α must be different from X_β for every ordinal $\beta <_{\mathbf{or}} \alpha$. If this condition is impossible to satisfy (because all the elements of Z have been assigned to ordinals less than α), no choice is made, and X_α is undefined.

More precisely, the above condition requires that the choice of the element X_α take place after the choices of all X_β, for all $\beta <_{\mathbf{or}} \alpha$. Still, we assume the choices are simultaneous, in the sense that they are activated at the same time for every ordinal in the universe. By activation we understand the command to execute the choice X_α for every ordinal α. The execution of the choice X_α is delayed until all the choices X_β have been executed for every $\beta <_{\mathbf{or}} \alpha$, and eventually it may be impossible to execute, so X_α is undefined.

Remark 8.1 We take the notion of free choice from intuitionistic mathematics, although, in general, we do not assume the usual restrictions associated with intuitionism. In particular, we do not assume that in dealing with free choice sequences we must impose continuity restrictions, as is usually the case in the frame of intuitionistic mathematics. On the other hand, we assume intuitionistic logical restrictions in dealing with non-local quantifiers. (see Chapter 10). □

The requirement that all choices should be made simultaneously, with the above restriction, is crucial, for we cannot think of a process where the ordinals are generated one after the other, although this is feasible, of course, if we consider only the natural numbers. Note that the whole construction depends essentially on the well-order properties of the ordinals. If we replace the ordinals with a different structure we may find that no permutation can be generated. For example, assume that for every natural number n in ω we associate a choice element X_{n} in the set Z, under the condition that for every $\mathsf{m} > \mathsf{n}$ a choice element X_{m} has been selected (so we are reversing the natural ordering in ω). It follows that no choice gets to be executed and no permutation is generated.

Even if we restrict the choice to ordinals, there are still some difficulties, for we intend to have the completion of an **FCTP** (in the same sense as the completion of the fundamental sequence in Chapter 7) to be a set in the universe, and this does not seem to be feasible since the transfinite permutation involves the collection of all ordinals, which is not a set. To overcome this difficulty we must refine our analysis, noting that although the ordinals are not a set, all choices take place in a set Z. In fact, we think it is reasonable to impose the following principle on the universe of sets.

Principle of Closure for Transfinite Permutations *In every free choice transfinite permutation over a set Z there is some ordinal where the choice is undefined.*

From the principle of closure we conclude that for every **FCTP** there is a least closure ordinal α where every element of Z is a choice of the form X_β for $\beta <_{\text{or}} \alpha$. We take as the completion of the permutation the set $F = \{<\beta, X_\beta> : \beta <_{\text{or}} \alpha :\}$. In fact, this F is an ordinal permutation of Z in the sense of Definition 8.1.1. So, it can be said that the completion of an **FCTP** produces a retraction, where a potentially universal object becomes an element of the universe. On the other hand, the completion of the fundamental sequence in Chapter 7 produces an expansion, where an infinite object is derived from a process that generates only finite sets.

Remark 8.2 The principle of closure means that, given an **FCTP**, and no matter the choices, a closure ordinal is reached where all the elements of Z have been generated. We conclude that there is no choice for the closure ordinal, and in fact there is no choice for the ordinals after the closure ordinal. We think this principle is reasonable, for otherwise we would have that Z contains a subset that is isomorphic to the ordinals, and this would mean that in some sense the set Z contains the universe. Note that the principle of closure belongs only to the informal theory of sets. In fact, the notion of **FCTP** is not a part of the formal theory. □

The idea of a free choice transfinite permutation over a set Z is critical, but it is not sufficient. The problem is that such objects are essentially abstractions, for the elements depend on choices that cannot be determined. We propose that a free choice permutation must be considered as an element of a larger totality, and can be characterized only as being a part of such a totality. So, rather than introducing individual free choice permutations, the rule we propose introduces a set that contains all possible free choice permutations.

To implement this general construction we consider a process where all the transfinite permutations over Z are generated at the same time. Here again, we activate all the ordinals simultaneously, under the constraint that action on the ordinal α requires previous action on every ordinal $\beta <_{\mathbf{or}} \alpha$. The action on α is a choice from Z for every permutation that is active at the time of the action.

The following example will clarify the intended meaning of the construction. Assume $Z = \omega$ (as a set), and consider the action at ω (as an ordinal). Infinitely many partial permutations have been generated at this time, for example:

$A = (0, 1, 2, 3, ...)$

$B = (1, 2, 3, ...)$

$C = (0, 2, 4, ...)$.

There is no action on A, because the set ω has been generated, so ω is not defined on A. There is only one possible action on B with the choice 0 that induces the new permutation $B' = (1, 2, 3, ..., 0)$. On the other hand, there are infinitely many possible actions on C where the possible choices are 1, 3, 5, etc.

We call this process *general free choice transfinite permutations over* Z, and the completion of the process is the set of all **FCTP** over Z. We denote this set by $\Xi(Z)$, and the rule below introduces the unary set operation Ξ in the system $\mathbb{G}$. First, we must extend the principle of closure to the process of general permutations.

Principle of Closure for General Transfinite Permutations *In the general free choice transfinite permutations over the set* Z *there is an ordinal where all the choices are undefined.*

As the restricted principle (see Remark 8.2), the general principle is derived from the locality of the set Z. Failure of the general principle would mean that every ordinal in the universe can be obtained as a permutation of the set Z, and this is a universal property that is not admissible for the sets in the universe. Essentially, we are postulating that the universe of sets is beyond any possible description in terms of a fixed local set.

The Ordinal Permutation Rule We introduce a unary set operation Ξ, which is characterized by the following axioms.

The Consistency Axiom:

$$(\forall F \in \Xi(Z))\mathbf{PM}(Z, F).$$

The Completeness Axiom: If p is a general binary set predicate, there is $F \in \Xi(Z)$ such that $\mathbf{do}(Z) = \alpha$ and

$$(\forall \beta \in \alpha)((\exists Y \in Z \setminus F[\beta])p(F{\upharpoonright}\beta, Y) \rightarrow p(F{\upharpoonright}\beta, \mathbf{ap}(F, \beta))).$$

In applications the predicate p is called the *choice predicate* of the axiom. Note that the completeness axiom asserts the existence of a set $F \in \Xi(Z)$, and by the consistency axiom this F is an ordinal permutation in the sense of Definition 8.1.1.

Remark 8.3 The description of the completeness axiom is a bit informal. Rather than a general binary predicate p, we must invoke a basic $(k+2)$-ary predicate p', such that with $\mathbf{X}$ fixed we have $p(V, Y) \equiv p'(V, Y, \mathbf{X})$. In this way the axiom assert the existence of $F \in \Xi(Z)$ that depends on $Z, \mathbf{X}$. More precisely, for every $(k+2)$-ary basic predicate p' there is one specific axiom, as described in the rule. In fact, the formal version of the axiom in rule **ST 14** in Chapter 9 is written in this way. The function F asserted in the axiom is said to be *induced* by the predicate p. If $p(F \restriction \beta, \mathbf{ap}(F, \beta))$ holds for some $\beta \in \mathbf{do}(F)$, we say that F *satisfies* p *at* β. If F does not satisfy p at β, we say that F *fails* p *at* β. □

EXERCISE 8.1 Assume the permutation $F \in \Xi(Z)$ is induced by the predicate p, and F fails p at β. Prove that if $Y \in Z$ and $p(F \restriction \beta, Y)$ holds, then $(\exists \gamma <_{\mathbf{or}} \beta)\mathbf{ap}(F, \gamma) = Y$.

Remark 8.4 From the completeness axiom, with p arbitrary, it follows that there is an ordinal permutation for every set Z. This is, of course, one of the standard formulations of the axiom of choice. On the other hand, we may have applications of the axiom where the function F is completely determined and no choice is implied. □

Theorem 8.1 *Let F and Z be sets. The following conditions are equivalent:*

(i) $F \in \Xi(Z)$.

(ii) $\mathbf{PM}(Z, F)$.

PROOF. The implication from (i) to (ii) is the consistency axiom. To prove the converse, assume F is an ordinal permutation of Z, and let $F' \in \Xi(Z)$

be induced by the choice predicate $p(V,Y) \equiv \mathbf{ap}(F,\mathbf{do}(V)) = Y$. By ordinal induction on $\beta \in \mathbf{do}(F)$ we show that $\beta \in \mathbf{do}(F')$ and $\mathbf{ap}(F,\beta) = \mathbf{ap}(F',\beta)$. It follows that $\mathbf{do}(F) = \mathbf{do}(F')$, hence $F = F'$. □

We prove two more consequences of the completeness axiom that are known to be equivalent to the axiom of choice.

Theorem 8.2 *Assume the set G is a function. There is a 1–1 function $F \subseteq G$ such that* $\mathbf{ra}(F) = \mathbf{ra}(G)$.

PROOF. If R is a 1–1 function we take $F = G$. Otherwise, let p be the binary predicate,

$$p(V,Y) \equiv (\forall \beta \in \mathbf{do}(V))[\mathbf{ap}(V,\beta)]_2 \neq [Y]_2,$$

and let F' be an ordinal permutation of G induced by the predicate p. From the assumption it follows that there is a least $\alpha <_{\mathbf{or}} \mathbf{do}(F')$ where the selection condition fails, so there is $\beta <_{\mathbf{or}} \alpha$ such that $[\mathbf{ap}(F',\beta)]_2 = [\mathbf{ap}(G,\alpha)]_2$. Hence we can take $F = F'[\alpha]$. □

Theorem 8.3 *Let X be a set such that $\bigcup X \neq \emptyset$. There is a function F such that* $\mathbf{do}(F) = X$ *and whenever $Y \in X$ and $Y \neq \emptyset$, then* $\mathbf{ap}(F,Y) \in Y$.

PROOF. Let G be a fixed ordinal permutation of X where $\mathbf{do}(G) = \alpha$ and using the completeness axiom introduce an ordinal permutation G' of $X \times \bigcup X$ such that if $\beta <_{\mathbf{or}} \alpha$, then $\mathbf{ap}(G',\beta) = \langle Y, Z \rangle$, where $\mathbf{ap}(G,\beta) = Y$ and in case $Y \neq \emptyset$, then $Z \in Y$. Hence we can take $F = G'[\alpha]$. □

EXERCISE 8.2 Define explicitly the choice predicate p that induces the ordinal permutation G' in the proof of Theorem 8.3.

8.2 Local Induction Revisited

Local induction was approached in Chapter 7 via the minimal fixed point of a monotonic operation satisfying some continuity assumptions. Here we show that the continuity assumptions can be avoided, using the ordinal permutation rule (or equivalently, using the power-set operation defined in Section 8.3).

8.2.1 $\mathbf{we}(X) = \{\mathbf{do}(F) : F \in \Xi(X) :\}$ (*the well-orders of* X).

Clearly, $\mathbf{we}(X)$ is a nonempty set of ordinals.

Theorem 8.4 *Let f be a unary set operation and Z a set. There is $\kappa \in \mathbf{we}(Z)$ and $\alpha \leq_{\mathbf{or}} \kappa$ such that $Z \cap f(\alpha) \subseteq \bigcup\{f(\gamma) : \gamma <_{\mathbf{or}} \alpha :\}$.*

PROOF. Let F be an ordinal permutation of Z induced by the predicate p such that

$$p(V, Y) \equiv Y \in f(\mathbf{do}(V)) \wedge Y \notin \bigcup\{f(\gamma) : \gamma \in \mathbf{do}(V) :\}.$$

We set $\mathbf{do}(F) = \kappa$, so $\kappa \in \mathbf{we}(Z)$. If the permutation F satisfies the predicate p at every $\gamma <_{\mathbf{or}} \kappa$, it follows that $Z \subseteq \bigcup\{f(\gamma) : \gamma <_{\mathbf{or}} \kappa :\}$, so we take $\alpha = \kappa$. Otherwise, there is a minimal ordinal $\alpha <_{\mathbf{or}} \kappa$ such that F fails p at α. It follows that $Z \cap f(\alpha) \subseteq \bigcup\{f(\gamma) : \gamma <_{\mathbf{or}} \alpha :\}$. □

EXERCISE 8.3 Let f be a given unary set operation and Z a set. Prove that there is ordinal $\kappa \in \mathbf{we}(Z)$ and ordinal $\alpha \leq_{\mathbf{or}} \kappa$ such that either $f(\alpha) \notin Z$, or there is $\gamma <_{\mathbf{or}} \alpha$ such that $f(\alpha) = f(\gamma)$.

Let Z be a set and g a unary set operation. We say that g is *monotonic on* Z if whenever X and Y are sets such that $X \subseteq Y \subseteq Z$, then $g(X) \subseteq g(Y) \subseteq Z$. This implies that whenever $X \subseteq Z$, then $g(X) \subseteq Z$. If $g(X) \subseteq X \subseteq Z$, we say that g *is closed at X in Z*. Clearly, if g is monotonic on Z, then g is closed at Z in Z. If $X \subseteq g(X) \subseteq Z$, we say that g *is reflexive at X in Z*. If g is closed and reflexive at X in Z, we say that X *is a fixed point of g* in Z. A *minimal fixed point of g* in Z is a fixed point X of g in Z such that whenever X' is another fixed point of g in Z, then $X \subseteq X'$.

EXERCISE 8.4 Prove that there is at most one set X that is a minimal fixed point of g in Z.

We take the operation $g^{\odot}$ as defined in 7.2.1. The next results are simply re-statements of results in Section 7.2.

Theorem 8.5 *If g is monotonic on Z, α is an ordinal, and Y,X are sets such that $Y \subseteq X \subseteq Z$, and g is closed at X in Z, then $g^{\odot}(\alpha, Y) \subseteq X$.*

PROOF. Same proof as in Theorem 7.7. □

Theorem 8.6 *Assume the operation g is monotonic on Z, α is an ordinal, and g is reflexive at Y in Z. Then:*

(i) $g^{\odot}(\alpha, Y) \subseteq Z$.

(ii) *g is reflexive at $g^{\odot}(\alpha, Y)$ in Z.*

(iii) *If $\gamma <_{\mathbf{or}} \alpha$, then $g^{\odot}(\gamma, Y) \subseteq g^{\odot}(\alpha, Y)$.*

(iv) $Y \subseteq g^{\odot}(\alpha, Y)$.

PROOF. Part (i) follows from Theorem 8.5 with $X = Z$. The proofs of (ii), (iii) and (iv) are as in Theorem 7.8, using part (i). □

EXERCISE 8.5 Assume g is monotonic on Z, α is a limit ordinal, and g is reflexive at Y in Z. Prove that $g^{\odot}(\alpha, Y) = \bigcup\{g^{\odot}(S(\gamma), Y) : \gamma <_{\mathbf{or}} \alpha :\}$.

Corollary 8.6.1 *Assume g is monotonic on Z, g is reflexive at Y in Z, and κ is an ordinal such that $g^{\odot}(\kappa, Y)$ is a fixed point of g. Then:*

(i) *If $\kappa <_{\mathbf{or}} \alpha$, then $g^{\odot}(\kappa, Y) = g^{\odot}(\alpha, Y)$.*

(ii) *If X is a fixed point of g and $Y \subseteq X \subseteq Z$, then $g^{\odot}(\kappa, Y) \subseteq X$.*

PROOF. Same proof as in Theorem 7.9. □

Corollary 8.6.2 *Assume g is monotonic on Z and g is reflexive at Y in Z. There is ordinal $\kappa \in \mathbf{we}(Z)$ and $\alpha \leq_{\mathbf{or}} \kappa$ such that $g^{\odot}(\alpha, Y)$ is a fixed point of g.*

PROOF. We apply Theorem 8.4 with $f(\alpha) = g^{\odot}(S(\alpha), Y)$ and Y fixed. It follows that there is ordinal $\kappa <_{\mathbf{or}} \mathbf{we}(Z)$ and $\alpha \leq_{\mathbf{or}} \kappa$ such that $g(g^{\odot}(\alpha, Y))$ $= g^{\odot}(S(\alpha), Y) \subseteq \{g^{\odot}(S(\gamma), Y) : \gamma <_{\mathbf{or}} \alpha :\} = g^{\odot}(\alpha, Y)$. By Theorem 8.6 (ii) it follows that $g^{\odot}(\alpha, Y)$ is a fixed point of g. □

Remark 8.5 The proof of the existence of the fixed point $g^{\odot}(\alpha, Y)$ in Corollary 8.6.2 (derived from Theorem 8.4) depends on the existence of the set Z and the monotonicity of g on Z. For example, the operation **pw** in 8.3.1 is monotonic in general (for $X \subseteq Y$ implies $\mathbf{pw}(X) \subseteq \mathbf{pw}(Y)$), but it is not monotonic on any set Z (for $\mathbf{pw}(Z) \not\subseteq Z$) and has no fixed point. On the other hand, the minimal fixed point itself is independent of Z, for the set operation $g^{\odot}$ does not depend on Z. □

Remark 8.6 In applications of Corollary 8.6.2 we are given a set Z and a set operation g monotonic on Z. From the theorem it follows that there is a set $\mathsf{Ind} = g^{\odot}(\alpha, \emptyset) \subseteq Z$ (the minimal fixed point of g in Z) that satisfies the following conditions:

(i) $g(\mathsf{Ind}) = \mathsf{Ind}$.

(ii) If $g(X) \subseteq X \subseteq Z$, then $\mathsf{Ind} \subseteq X$.

These conditions induce a method of proof by Ind-*induction*, where in order to prove $\mathsf{Ind} \subseteq X$ we prove that $g(X) \subseteq X$. Here, the induction hypothesis is $Y \in g(X)$, which is used to prove $Y \in X$. □

EXERCISE 8.6 Assume g is monotonic on Z and the set $Y \subseteq Z$ satisfies the condition $Y \subseteq g(Y)$. Prove that there is a set $X_0 \subseteq Z$ such that:

(a) X_0 is a fixed point of g in Z and $Y \subseteq X_0$.

(b) If $Y \subseteq X' \subseteq Z$ and g is closed at X' in Z, then $X_0 \subseteq X'$.

Remark 8.7 Applications of local induction can be derived from predicates satisfying some monotonic conditions. If p is a binary predicate, we say that p is *monotonic on* Z if whenever $X \subseteq X' \subseteq Z$, $Y \in Z$ and $p(X, Y)$ holds, then $p(X', Y)$ also holds. We can define from p the unary set operation $g(X) = \{Y \in Z :: p(X, Y)\}$, that it is also monotonic on Z. If Ind is the minimal fixed point of g, then the following conditions are satisfied:

(i) If $Y \in Z$, then $p(\mathsf{Ind}, Y) \equiv Y \in \mathsf{Ind}$.

(ii) If $X \subseteq Z$ and $Y \in Z$ whenever $p(X, Y)$, then $\mathsf{Ind} \subseteq X$.

Here again we have a method of proof by induction, where in order to prove $\mathsf{Ind} \subseteq X$ we prove $(\forall Y \in Z)(p(X, Y) \rightarrow Y \in X)$. This means that from the induction hypothesis $p(X, Y)$ we must derive $Y \in X$.

Typical applications of inductions derived from monotonic predicates take place in mathematical logic, where the predicate $p(X, Y)$ describes the relation where (the assertion) Y is a *logical consequence* of the set (of assertions) X, relative to some system of logic. In this case Ind is the set of all *theorems* of the system. □

EXERCISE 8.7 Prove conditions (i) and (ii) in Remark 8.7.

EXAMPLE 8.1 Consider the binary predicate $p(X, Y) \equiv Y = 1 \vee (\exists \mathsf{n} \in X) Y = \mathsf{n} \times_{\mathsf{or}} 2$. Clearly, p is monotonic on ω, so the construction in Remark 8.8 defines a set $\mathsf{Ind} \subseteq \omega$. The elements of Ind are all the powers of 2. □

8.3 Power-Set

From the ordinal permutation rule we can derive several classical constructions in standard set theory. We have already discussed the power-set of a finite set (Theorem 4.6) and the set of all finite subsets of an arbitrary set X (Definition 7.1.1). Here we consider the general power-set construction.

8.3.1 $\mathbf{pw}(X) = \bigcup\{\{F[\beta] : \beta \leq_{\mathbf{or}} \mathbf{do}(F) :\} : F \in \Xi(X) :\}$ (*the power set of* X).

Note that in this definition we are using the replacement rule twice (in both cases under rule **BT 7**) and, furthermore, general union.

EXAMPLE 8.2 Let W be a subset of ω. Then there is $F \in \Xi(\omega)$ such that $\mathbf{do}(F) = \alpha \leq_{\mathbf{or}} \omega + \omega$, and $W = F[\beta]$ for some $\beta \leq_{\mathbf{or}} \alpha$. If W is finite we can use the counting of W, extended by generating in increasing order the elements of $W' = \omega \setminus W$. In this case $\alpha = \omega$ and $\beta <_{\mathbf{or}} \omega$. If W is infinite we can generate the elements of W in increasing order and then the elements of W' also in increasing order. In this case $\omega = \beta \leq_{\mathbf{or}} \alpha \leq_{\mathbf{or}} \omega +_{\mathbf{or}} \omega$. Note that in both cases the function F is completely determined and the completeness axiom is not required. □

Theorem 8.7 *Let X and W be sets. The following conditions are equivalent:*

(i) $W \in \mathbf{pw}(X)$.

(ii) $W \subseteq X$.

PROOF. Assume (i) holds. Then $W = F[\beta]$ where $F \in \Xi(X)$ and $\beta \leq \mathbf{do}(F)$. It follows that $W \subseteq X$. Assume now that W is a subset of X and introduce the choice predicate,

$$p(V, Y) \equiv Y \in W.$$

Let F be an ordinal permutation of X induced by the completeness axiom, where $\mathbf{do}(F) = \alpha$. If $W = X$ we have $F[\alpha] = W$. Otherwise, there is a least $\beta <_{\mathbf{or}} \alpha$ such that $\mathbf{ap}(F, \beta) \notin W$. Clearly, $F[\beta] = W$. In either case we have $W \in \mathbf{pw}(X)$. □

Remark 8.8 The proof of Theorem 8.7 involves the axiom of choice via the completeness axiom of the ordinal permutation rule. This may not be necessary in some cases. For example, if $X = \omega$, the argument of Example 8.2 shows that the function F is completely determined without choices. □

EXERCISE 8.8 Let g, Z, Y, and α be as in Corollary 8.6.2. Prove that $g(\alpha, Y) = \bigcap\{V \in \mathbf{pw}(Z) :: g(V) \subseteq V \wedge Y \subseteq V\}$.

EXERCISE 8.9 Prove that the predicate $\subseteq$ is extensional in the sense of Section 2.2.

EXERCISE 8.10 Let X and Y be sets. Prove that $X \subseteq Y$ if and only if $\mathbf{pw}(X) \subseteq \mathbf{pw}(Y)$.

EXERCISE 8.11 Prove that if the set X is transitive, then $\mathbf{pw}(X)$ is also transitive.

EXERCISE 8.12 Let X be a well-founded set and Y a set such that $\mathbf{pw}(Y) \in X$. Prove:

(a) $Y \in \bigcup X$.

(b) $\bigcup Y \subseteq \bigcup\bigcup\bigcup X$.

(c) $\mathbf{pw}(X) \notin Y$.

EXERCISE 8.13 Let X be a well-founded set and Y a set such that $X \cup \{\mathbf{pw}(X)\} \subseteq Y \cup \{\mathbf{pw}(Y)\}$. Prove that $X \subseteq Y$ and either $X = Y$ or $\mathbf{pw}(X) \in Y$.

EXERCISE 8.14 Prove that if the set X is well-founded, then $\mathbf{pw}(X)$ is also well-founded.

Using ordinal recursion we can iterate the power-set construction. The result is the so-called von Neumann *cumulative hierarchy*. In fact, this iteration can be derived from the general iteration given by Definition 7.2.1

8.3.2 $\mathbf{pw}^*(\alpha) = \mathbf{pw}^{\odot}(\alpha, \emptyset)$.

Note that the operation **pw** is monotonic, hence Theorems 7.7 and 7.8 are valid. On the other hand, the operation **pw** is not continuous in the sense of Remark 7.3, and it has no fixed point under Theorem 7.10. Similarly, it has no fixed point under Corollary 8.6.2, because there is not set Z such that **pw** is monotonic on Z. In fact, it can be proved that there is no set X in the universe, such that $\mathbf{pw}(X) = X$.

EXERCISE 8.15 Let α be an arbitrary ordinal. Prove that $\mathbf{pw}^*(\alpha)$ is transitive and well-founded.

EXERCISE 8.16 Assume $\beta <_{\mathbf{or}} \alpha$. Prove $\mathbf{pw}^*(\beta) \in \mathbf{pw}^*(\alpha)$ and $\mathbf{pw}^*(\beta) \subset \mathbf{pw}^*(\alpha)$

EXERCISE 8.17 Let X be a well-founded set where $\mathbf{rk}(X) = \alpha$. Prove that $X \in \mathbf{pw}^*(\alpha)$.

EXERCISE 8.18 Prove that if X is a set, then $\mathbf{pw}(X) \not\subseteq X$.

EXERCISE 8.19 Let α be an ordinal. Prove that $\alpha \subset \mathbf{pw}(\alpha)$.

8.4 Cardinals

We derive the classical theory of cardinals from the operation $\mathbf{we}(Z)$. First, we prove several general properties.

Theorem 8.8 *Assume X is a set, $\alpha \in \mathbf{we}(X)$ and $\beta \in \mathbf{we}(\mathbf{pw}(X))$. Then $\alpha <_{\mathbf{or}} \beta$.*

PROOF. Assume $F_1 \in \Xi(X)$ with $\mathbf{do}(F_1) = \alpha$ and $F_2 \in \Xi(\mathbf{pw}(X))$ with $\mathbf{do}(F_2) = \beta$. We introduce a set,

$$A = \{Y \in X :: (\exists\gamma <_{\mathbf{or}} \beta)(\mathbf{ap}(F_1, \gamma) = Y \wedge Y \notin \mathbf{ap}(F_2, \gamma))\}.$$

Since $A \in \mathbf{pw}(X)$ it follows that there is $\gamma <_{\mathbf{or}} \beta$ such that $\mathbf{ap}(F_2, \gamma) = A$. To get a contradiction we assume $\beta \leq_{\mathbf{or}} \alpha$, hence there is $Y \in X$ such that $\mathbf{ap}(F_1, \gamma) = Y$. From these assumptions it follows that $Y \in A$ if and only if $Y \notin A$, and this is a contradiction. We conclude that $\alpha <_{\mathbf{or}} \beta$. □

Theorem 8.9 *Assume $X \subset Z$. Then:*

(i) *If $\alpha \in \mathbf{we}(X)$, then there is $\beta \in \mathbf{we}(Z)$ such that $\alpha <_{\mathbf{or}} \beta$.*

(ii) *If $\beta \in \mathbf{we}(Z)$, then there is $\alpha \in \mathbf{we}(X)$ such that $\alpha \leq_{\mathbf{or}} \beta$.*

PROOF. To prove (i) assume $F \in \Xi(X)$ with $\mathbf{do}(X) = \alpha$. Introduce a general predicate p such that

$$p(V, Y) \equiv Y = \mathbf{ap}(F, \mathbf{do}(V)),$$

and let F' be the ordinal permutation of Z induced by p as a choice predicate. If $\mathbf{do}(F') = \beta$, and $\beta \leq_{\mathbf{or}} \alpha$, it follows that $F = F' \upharpoonright \beta$, hence $Z \subseteq X$, and this is a contradiction. We conclude that $\alpha <_{\mathbf{or}} \beta$.

To prove (ii) assume $F \in \Xi(Z)$ with $\mathbf{do}(F) = \beta$. We set $A = \{\gamma \in \beta :: \mathbf{ap}(F, \gamma) \in X\}$. The set operation **oc** (definition 6.6.1) enumerates the ordinals in A in increasing order, so we set

$$F' = \{<\gamma, \mathbf{ap}(F, \mathbf{oc}(\gamma, A))> : \gamma <_{\mathbf{or}} \mathbf{oc}'(A) :\}.$$

It follows that $F' \in \Xi(X)$ and $\mathbf{do}(F') = \mathbf{oc}'(A) \leq_{\mathbf{or}} \beta$. □

Theorem 8.10 *Assume α, β, and γ are ordinals such that $\alpha <_{\mathbf{or}} \beta <_{\mathbf{or}} \gamma$. Assume also that X is a set, $\alpha \in \mathbf{we}(X)$, and $\gamma \in \mathbf{we}(X)$. Then $\beta \in \mathbf{we}(X)$.*

PROOF. From the assumptions we know there is an ordinal permutation F_1 of X, where $\mathbf{do}(F_1) = \alpha$ and there is also an ordinal permutation F_2 of X where $\mathbf{do}(F_2) = \gamma$. Using inversion and composition we get a 1–1 function F_3 where $\mathbf{do}(F_3) = \gamma$ and $\mathbf{ra}(F_3) = \alpha$. We can apply the Cantor Bernstein Theorem (Example 7.2) to F_3 and get a 1–1 function F_4 where $\mathbf{do}(F_4) = \gamma$

and $\mathbf{ra}(F_4) = \beta$. Finally, using inversion and composition with F_2, we get a 1–1 function F where $\mathbf{do}(F) = \beta$ and $\mathbf{ra}(F) = X$. Clearly, $F \in \Xi(X)$, so it follows that $\beta \in \mathbf{we}(X)$. □

The theory of cardinals can be derived from the following two definitions.

8.4.1 $\mathbf{cd}(X) = \mathbf{inf}(\mathbf{we}(X))$ (*the cardinal of X*).

8.4.2 $\mathbf{ncd}(X) = \mathbf{sup}^+(\mathbf{we}(X))$ (*the next cardinal after X*).

Theorem 8.11 *Let X be a set, and α an ordinal. Then:*

(i) $\mathbf{cd}(X)$ *and* $\mathbf{ncd}(X)$ *are ordinals.*

(ii) $\mathbf{cd}(X) \in \mathbf{we}(X)$.

(iii) *If* $\alpha \in \mathbf{we}(X)$, *then* $\mathbf{cd}(X) \leq_{\mathbf{or}} \alpha <_{\mathbf{or}} \mathbf{ncd}(X)$.

(iv) $\mathbf{ncd}(X) \notin \mathbf{we}(X)$.

(v) *If* $\mathbf{cd}(X) \leq_{\mathbf{or}} \alpha <_{\mathbf{or}} \mathbf{ncd}(X)$, *then* $\alpha \in \mathbf{we}(X)$.

PROOF. Part (i) follows from Theorem 6.3 (iv) and Theorem 6.1 (v). Part (ii) follows from Theorem 6.3 (iii). Part (iii) follows from Theorem 6.3 (i) and Theorem 6.5 (ii). Part (iv) follows from (iii). To prove (v) note that from Theorem 6.5 (i) it follows that there is $\alpha' \in \mathbf{we}(X)$ such that $\alpha \leq_{\mathbf{or}} \alpha'$, so by Theorem 8.10 we have $\alpha \in \mathbf{we}(X)$. □

Corollary 8.11.1 *If X is a finite set, then* $\mathbf{we}(X) = \{\mathbf{ct}(X)\}$, $\mathbf{cd}(X) = \mathbf{ct}(X)$ *and* $\mathbf{ncd}(X) = S(\mathbf{ct}(X))$.

PROOF. If $F \in \Xi(X)$, then by Corollary 5.6.1 it follows that $\mathbf{do}(F) = \mathbf{ct}(X)$, hence $\mathbf{we}(X) = \{\mathbf{ct}(X)\}$. The rest follows from Corollary 6.5.1. □

Corollary 8.11.2 *Assume X is infinite and $\alpha \in \mathbf{we}(X)$. Then:*

(i) *α is an infinite ordinal.*

(ii) $S(\alpha) \in \mathbf{we}(X)$.

(iii) $\mathbf{ncd}(X) = \mathbf{sup}(\mathbf{we}(Z))$.

PROOF. We know that $\alpha = \mathbf{do}(F)$ where $F \in \Xi(X)$. So α cannot be finite, for by Corollary 4.1.4 this implies that X is finite. To prove (ii) we note that $\omega \leq_{\mathrm{or}} \alpha$ and define an ordinal permutation F', where $\mathbf{do}(F') = S(\alpha)$, as follows: If $\beta \leq_{\mathrm{or}} \alpha$ is a natural number we set $\mathbf{ap}(F', \beta) = \mathbf{ap}(F, S(\beta))$. If $\omega \leq_{\mathrm{or}} \alpha$ we set $\mathbf{ap}(F', \beta) = \mathbf{ap}(F, \beta)$. Finally, we set $\mathbf{ap}(F', \alpha) = \mathbf{ap}(F, 0)$. Clearly, $F' \in \Xi(X)$. Finally, part (iii) follows from Corollary 6.5.1 (ii), noting that by (ii) the condition $\mathbf{sup}(\mathbf{we}(X)) \in \mathbf{we}(X)$ is impossible. □

EXERCISE 8.20 Let X be a set and α an ordinal. Prove the following conditions are equivalent:

(a) $\mathbf{cd}(X) = \alpha$.

(b) $\alpha \in \mathbf{we}(X) \wedge (\forall \beta <_{\mathrm{or}} \alpha)\beta \notin \mathbf{we}(X)$.

(c) $\alpha \in \mathbf{we}(X) \wedge (\forall \beta \in \mathbf{we}(X))\alpha \leq_{\mathrm{or}} \beta)$.

EXERCISE 8.21 Let X be a set. Prove that $\mathbf{cd}(X) <_{\mathrm{or}} \mathbf{cd}(\mathbf{pw}(X))$.

EXERCISE 8.22 Let X and Z be sets such that $X \subset Z$. Prove:

(a) $\mathbf{cd}(X) \leq_{\mathrm{or}} \mathbf{cd}(Z)$.

(b) $\mathbf{ncd}(X) \leq_{\mathrm{or}} \mathbf{ncd}(Z)$.

8.4.3 $\mathbf{CD}(X) \equiv \mathbf{cd}(X) = X$ (*X is a cardinal*).

8.4.4 $X \sim Y \equiv \mathbf{we}(X) = \mathbf{we}(Y)$ (*X is similar to Y*).

EXERCISE 8.23 Prove that ω is a cardinal.

EXERCISE 8.24 Let α be an ordinal. Prove that α is a cardinal if and only if $(\forall \beta <_{\mathrm{or}} \alpha)\neg(\beta \sim \alpha)$.

Theorem 8.12 *If X and Y are sets, then the following conditions are equivalent:*

(i) $X \sim Y$.

(ii) $\mathbf{cd}(X) = \mathbf{cd}(Y)$.

(iii) $\mathbf{we}(X) \cap \mathbf{we}(Y) \neq \emptyset$.

(iv) $(\exists F \in \mathbf{pw}(X \times Y))(\mathbf{FU1}(F) \wedge \mathbf{do}(F) = X \wedge \mathbf{ra}(F) = Y)$.

PROOF. The implication from (i) to (ii) is obvious. From (ii) we can prove (iii) noting that $\mathbf{cd}(X) \in \mathbf{we}(X)$ and $\mathbf{cd}(Y) \in \mathbf{we}(Y)$. The implication from (iii) to (iv) follows, using inversion and composition. Part (i) follows from (iv), using inversion and composition. □

Corollary 8.12.1 *Assume α is an ordinal and X is a set. Then $\alpha \sim X$ if and only if $\alpha \in \mathbf{we}(X)$.*

PROOF. We note that $\alpha \in \mathbf{we}(\alpha)$, hence the equivalence follows from the definition and Theorem 8.12 (iii). □

Corollary 8.12.2 *Assume α, β, and γ are ordinals such that $\alpha <_{\mathbf{or}} \beta <_{\mathbf{or}} \gamma$ and $\alpha \sim \gamma$. Then $\alpha \sim \beta \sim \gamma$.*

PROOF. Immediate from Theorem 8.10. □

EXERCISE 8.25 Let X and Y be sets. Prove that $X \sim Y$ if and only if $\mathbf{ncd}(X) = \mathbf{ncd}(Y)$.

EXERCISE 8.26 Let X and Y be sets, and assume F is a function where $\mathbf{do}(F) = X$ and $\mathbf{ra}(F) = Y$. Prove that $\mathbf{cd}(Y) \leq_{\mathbf{or}} \mathbf{cd}(X)$ and $\mathbf{ncd}(Y) \leq_{\mathbf{or}} \mathbf{ncd}(X)$.

There is a classical connection between the ordinal permutations and the rank theory of well-founded relations (see Corollary 6.9.1).

Theorem 8.13 *Let α be an ordinal and X a set. The following conditions are equivalent:*

(i) $(\exists R \in \mathbf{pw}(X \times X))(\mathbf{fd}(R) = X \wedge (\forall Z \in \mathbf{fd}(R))\mathbf{WF}^*(Z, R) \wedge \mathbf{rk}^{**}(R) = \alpha)$.

(ii) $\alpha \in \mathbf{we}(X)$.

PROOF. Assume (i) holds and set the function F as in Remark 5.2. It follows that F is 1–1, $\mathbf{do}(F) = Z$ and $\mathbf{ra}(F) = \alpha$, so $\alpha \in \mathbf{we}(X)$. If (ii) holds we define R as in Example 6.4. □

Theorem 8.14 *Let X be a set. Then:*

(i) $\mathbf{cd}(X)$ *is a cardinal.*

(ii) $\mathbf{ncd}(X)$ *is a cardinal.*

(iii) *If* $\mathbf{cd}(X) <_{\mathbf{or}} \alpha <_{\mathbf{or}} \mathbf{ncd}(X)$ *then α is not a cardinal.*

PROOF. To prove (i) we set $\mathbf{cd}(X) = \alpha$, hence $\alpha \sim X$, and $\mathbf{cd}(\alpha) = \mathbf{cd}(X) = \alpha$. To prove (ii) we set $\mathbf{cd}(\mathbf{ncd}(X)) = \alpha$ and note that $\alpha \leq_{\mathbf{or}} \mathbf{ncd}(X)$. If $\alpha <_{\mathbf{or}} \mathbf{ncd}(X)$ there is $\beta \in \mathbf{we}(X)$ such that $\alpha \leq_{\mathbf{or}} \beta <_{\mathbf{or}} \mathbf{ncd}(X)$, and by Corollary 8.12.2 it follows that $\beta \sim \mathbf{ncd}(X)$, which means that $\mathbf{ncd}(X) \in \mathbf{we}(X)$, which contradicts Theorem 8.11 (iv). To prove part (iii) we note that from the assumption and Theorem 8.11 (v) it follows that $\alpha \in \mathbf{we}(X)$, hence $\mathbf{cd}(\alpha) = \mathbf{cd}(X) <_{\mathbf{or}} \alpha$. □

Theorem 8.15 *Let α and β be ordinals. Then:*

(i) *If* $\alpha <_{\mathbf{or}} \beta$, *then* $\mathbf{cd}(\alpha) \leq_{\mathbf{or}} \mathbf{cd}(\beta)$.

(ii) *If* $\mathbf{cd}(\alpha) <_{\mathbf{or}} \mathbf{cd}(\beta)$, *then* $\alpha <_{\mathbf{or}} \beta$.

PROOF. Assume $\alpha <_{\mathbf{or}} \beta$. Since $\mathbf{cd}(\beta) \leq_{\mathbf{or}} \beta$ there are two possibilities: either $\mathbf{cd}(\beta) \leq_{\mathbf{or}} \alpha <_{\mathbf{or}} \beta$, hence $\mathbf{cd}(\alpha) = \mathbf{cd}(\beta)$ by Theorem 8.11 (v), or $\mathbf{cd}(\alpha) <_{\mathbf{or}} \alpha <_{\mathbf{or}} \mathbf{cd}(\beta)$. Part (ii) follows immediately from (i). □

Corollary 8.15.1 *Let X be a set of cardinals. Then* $\mathbf{sup}(X)$ *is a cardinal.*

PROOF. We set $\mathbf{sup}(X) = \alpha$ and assume $\mathbf{cd}(\alpha) <_{\mathbf{or}} \alpha$. This means that $\mathbf{cd}(\alpha) <_{\mathbf{or}} \alpha' \leq_{\mathbf{or}} \alpha <_{\mathbf{or}} \mathbf{ncd}(\alpha)$ where $\alpha' \in X$. Since α' is a cardinal, this contradicts Theorem 8.14 (iii). □

The theory of cardinals in this chapter is extended in the Appendix via the enumeration rule that can be derived from the ordinal permutation rule.

8.5 Notes

The theory presented up to Chapter 7 of this monograph is a very restricted, but still significant, fragment of what is usually identified with full set theory. This fragment is clearly predicative, in the limited sense in which we use this term, namely, no construction involves a reference to the universe of sets. In particular, we have made a definite effort to show that the introduction of ω is completely determined by the fundamental sequence.

In this chapter we have tried to extend the predicative frame in order to produce a construction for the power-set and the theory of cardinals. There is no doubt that some form of extension beyond the omega rule is essential, if the theory is going to have a claim to a serious consideration, and the ordinal permutation rule provides a reasonable approach supported by a legitimate construction. On the other hand, we agree that there are some problems with this rule, and a better proposal would be desirable. Still, we think it is a better approach than the crude power-set axiom, because the axiom of choice contains constructive elements that provide a foundation for the completeness axiom. In fact, these elements were already obvious in the classical proof of Zermelo (1904). Similar concerns about the power-set axiom have been expressed in the literature, although usually in a low key tone (for example, see Levy(1979) page 99). At any rate, the ordinal permutation rule and the whole system $\mathbb{G}$ is consistent with Zermelo-Fraenkel plus the axiom of choice.

The permutation rule, as introduced in this chapter, provides an interesting example where the axiom of choice is introduced in an operational frame. Note that the proof of the power-set construction (Theorem 8.7) depends essentially on the choice dimension of the completeness axiom.

Another alternative, more conservative, is offered in the Appendix, where we introduce an operation $\Upsilon(Z, \alpha)$, consisting of all the α-enumerations of the set Z. Essentially, this construction gives the power-set of all well-ordered sets, in particular the power-set of ω.

The theory of local induction can be generalized by replacing the relation $\subseteq$ with a partial ordering relation R that satisfies several closure properties. For example, see the proof of Theorem 1.2 in Sanchis (1992), where the structure is called a *domain* and the set theoretical frame is not completely defined.

Chapter 9

Formalization: Classical Logic

In the formulation of our system of set theory we have chosen a partially formalized style consisting of rules for the introduction of predicates and operations. In this way the theory is expressed as a system of constructions, and this is of special relevance within the constructive predicative approach we support. Furthermore, from a practical point of view this style appears to be particularly useful, as sets are systematically generated by well understood constructions. We expect this type of presentation to be appealing to mathematicians interested in concrete applications of set theory.

We complete this work with a complete formalization of the theory within the frame of first order predicate logic with equality. This is not a trivial matter, since we want to preserve the operational structure of the system. This requires us to define the basic syntax of constants and formulas simultaneously with the definition of the axioms of the system. Furthermore, we must discuss the underlying logic, since we have anticipated that in the general theory we expect to rely on intuitionistic logic. On the other hand, there is no reason to avoid classical logic in the more restricted field of local formulas. We discuss the formalization of logic, both classical and intuitionistic, in the frame of the well-known calculus of sequents due to G. Gentzen. Here, and in the next chapter, we assume familiarity with the usual techniques of mathematical logic.

9.1 Syntax and Axioms

We assume a countable number of set variables, which we identify with the variables used in the text: $A, A', A_1, B, ...$ We assume also two initial binary predicate constants: **EP** (membership), and **EQ** (equality). We assume the following initial operation constants: $\emptyset$ (emptyset), **ins** (insertion), ω (omega), and Ξ (permutation).

We assume a countable number of predicate symbols, that may become predicate constants under the rules given below. In the same way, we assume a countable number of operation symbols that may become operation constants.

We proceed now to define by simultaneous induction *predicate constants, operation constants, terms, formulas, local formulas*, and *set axioms*. The definition involves the following rules:

ST 0 The initial predicate constants are predicate constants and the initial operation constants are operation constants.

ST 1 A set variable is a term.

ST 2 If p is a k-ary predicate constant, $k \geq 0$, and $V_1, ..., V_k$ are terms, then $p(V_1, ..., V_k)$ is a (local) formula.

ST 3 If f is a k-ary operation constant, $k \geq 0$, and $V_1, ..., V_k$ are terms, then $f(V_1, ..., V_k)$ is a term.

ST 4 If ϕ and ψ are (local) formulas, then $\neg\phi$, $\phi \vee \psi$, $\phi \wedge \psi$, and $\phi \rightarrow \psi$ are also (local) formulas.

ST 5 If ϕ is a (local) formula, V is a term, and X is a set variable that does not occur in V, then $(\forall X \in V)\phi$ and $(\exists X \in V)\phi$ are (local) formulas.

ST 6 If ϕ is a formula and X is a set variable, then $(\forall X)\phi$ and $(\exists X)\phi$ are formulas.

ST 7 The following formula is a set axiom:

$$(\forall Z \in X)(Z \in Y) \wedge (\forall Z \in Y)(Z \in X) \rightarrow X = Y.$$

ST 8 The formal axioms of the following rules are set axioms: emptyset, insertion, omega. The consistency axiom of the ordinal permutation rule is also a set axiom.

ST 9 If V is a term, the variables in V are in the list $\mathbf{X}$, and f is a new operation symbol, we can introduce f as a k-ary operation constant with the set axiom

$$f(\mathbf{X}) = V.$$

ST 10 If ϕ is a local formula, the free variables in ϕ are in the list $\mathbf{X}$, and p is a new predicate symbol, we can introduce p as a k-ary predicate constant with the set axiom:

$$p(\mathbf{X}) \equiv \phi.$$

ST 11 Let p be a $(k+s+1)$-ary predicate constant, $g_1,...,g_s$ be k-ary operation constants, and h a $(k+s)$-ary operation constant. If f is a new operation symbol we can introduce f as a k-ary operation constant with the set axiom

$$Z \in f(\mathbf{X}) \equiv (\exists Y_1 \in g_1(\mathbf{X}))...(\exists Y_s \in g_s(\mathbf{X}))(Z \in h(\mathbf{Y},\mathbf{X}) \wedge p(Z,\mathbf{Y},\mathbf{X})).$$

ST 12 Let q_1 and q_2 be $(k+1)$-ary predicate constants and g a $(k+1)$-ary operation constant. If p is a new predicate symbol and f is a new operation symbol, we can introduce p as a new $(k+1)$-ary predicate constant and f as a new $(k+1)$-ary operation constant with the axioms derived from the set induction rule in Chapter 3, where $p = \mathsf{ind}$ and $f = \mathbf{tc}_{\mathsf{ind}}$.

ST 13 If p and f are predicate and operation constants introduced by an application of rule **ST 12**, h is a $(k+2)$-ary operation constant, and f' is a new operation symbol, we can introduce f' as a new $(k+1)$-ary operation constant, with the set axiom

$$f'(Z,\mathbf{X}) = [p(Z,\mathbf{X}) \Rrightarrow h((\lambda Y \in f(Z,\mathbf{X}))f'(Y,\mathbf{X}), Z, \mathbf{X}), \emptyset].$$

ST 14 If p is a $(k+2)$-ary predicate constant, then the following is a set axiom:

$$(\exists F \in \Xi(Z))(\forall \beta \in \mathbf{do}(F))((\exists Y \in Z \setminus F[\beta])p(F{\restriction}\beta, Y, \mathbf{X}) \to p(F{\restriction}\beta, \mathbf{ap}(F,\beta), \mathbf{X})).$$

ST 15 If ϕ is a local formula, then $\phi \vee \neg\phi$ is a set axiom.

ST 16 The usual equality axioms: reflexivity, symmetry, transitivity, and replacement in predicate constants and operation constants.

The following comments should clarify some peculiarities of the rules.

1) The symbol $\equiv$ is assumed to be defined in terms of conjunction and conditional.

2) The translation of the axiom for the omega rule in **ST 8** requires the introduction of the predicate **NT** via rule **ST 12**.

3) The translation of the axioms for the ordinal permutation rule in **ST 8** and **ST 14** requires the introduction of several operations and predicates, in particular the predicate **OR** via rule **ST 12**.

4) In rule **ST 13** the lambda sub-term is generated using rule **ST 11**, as explained in Chapter 2.

5) All axioms are local formulas with free variables.

6) The distinction between free and bound variables is the usual one in mathematical logic. The only binding mechanism is induced by the local and global quantifiers. In particular, we do not include in this formalization the binding structure induced by rules **BT 6** and **BT 7** in Chapter 3.

7) The universal closure of a set axiom is called a *closed axiom*. The closed axioms are the real axioms of the theory.

8) The formulas introduced by rule **ST 2** are called *atomic formulas*.

9) Concerning the form of the axiom given by rule **ST 14**, see Remark 8.3.

EXERCISE 9.1 Let ϕ be a formula. Prove the following conditions are equivalent:

(a) ϕ is a local formula.

(b) ϕ contains no global quantifiers.

9.2 Sequents

In order to formalize the notion of logical deduction we use Gentzen rules for the calculus of sequents. In this way we can deal simultaneously with classical and intuitionistic logic.

Letters $\mathcal{M}, \mathcal{N}$ will be used to denote finite sequences of formulas, possibly empty. The formula ϕ is a *component* of the sequence $\mathcal{M}$ if $\mathcal{M}$ contains one or more occurrences of ϕ. We write $\mathcal{M} < \mathcal{N}$ to indicate that every component of $\mathcal{M}$ is also a component of $\mathcal{N}$ and say that $\mathcal{M}$ is *included* in $\mathcal{N}$. A sequence $\mathcal{M}$ of length n is *deterministic* if $n \leq 1$.

A pair $(\mathcal{M}, \mathcal{N})$ of sequences, which we shall write in the form $\mathcal{M} \vdash \mathcal{N}$, we call a *sequent*. If $\mathcal{N}$ is deterministic we say that the sequent $\mathcal{M} \vdash \mathcal{N}$ is *deterministic*. An *initial* sequent is a sequent of the form $\phi \vdash \phi$, where ϕ is an atomic formula.

We define a classical calculus of sequents where new sequents are derived from the initial sequents by means of the rules described below. An intuitionistic sub-calculus is derived later by restricting the rules to deterministic sequents.

If ϕ is a formula, X is a variable, and V is a term, then $\phi(X/V)$ denotes the result of substituting the term V for all the free occurrences of X in ϕ. As usual, we say that the substitution is *free* if no variable in V becomes bound after the substitution. This transformation can be iterated, so if X' is a variable and V' is a term we can write $\phi(X/V)(X'/V')$ which describes a process where first we perform the substitution of V for X, and next the substitution of V' for X'. Note that if X' occurs in V, the second substitution will change V to a different term.

Note that the variable X can be taken arbitrarily, for if Y is a new variable different from all variables in ϕ, then $\phi(X/V)$ is the same formula as $\phi(X/Y)(Y/V)$. The trivial substitution $\phi(X/X)$ is written simply $\phi(X)$.

The substitution notation can be extended to sequences of formulas, and also to sequents. If $\mathcal{M}$ is the sequence ϕ_1, ϕ_2, ϕ_3, then $\mathcal{M}(X/V)$ is the sequence $\phi_1(X/V), \phi_2(X/V), \phi_3(X/V)$. If S is the sequent $\mathcal{M} \vdash \mathcal{N}$, then $S(X/V)$ is the sequent $\mathcal{M}(X/V) \vdash \mathcal{N}(X/V)$. These substitutions are *free* if each of the individual substitutions is free.

The rules of the calculus are expressed using the following notation. A unary rule has the form $S \Rightarrow S'$, which means that the sequent S' can be derived from the sequent S. A binary rule has the form $S_1; S_2 \Rightarrow S'$, which means that the sequent S' can be derived from the sequents S_1 and S_2.

In the following rules the letters ϕ and χ denote formulas, the letters X and Y denote variables, and the letters V and W denote terms.

INC: $\mathcal{M} \vdash \mathcal{N}; \mathcal{M} < \mathcal{M}'; \mathcal{N} < \mathcal{N}' \Rightarrow \mathcal{M}' \vdash \mathcal{N}'$.

NG-R: $\phi, \mathcal{M} \vdash \mathcal{N} \Rightarrow \mathcal{M} \vdash \mathcal{N}, \neg\phi$.

NG-L: $\mathcal{M} \vdash \mathcal{N}, \phi \Rightarrow \neg\phi, \mathcal{M} \vdash \mathcal{N}$.

DJ-R: $(a)\mathcal{M} \vdash \mathcal{N}, \phi \Rightarrow \mathcal{M} \vdash \mathcal{N}, \phi \vee \chi$ - $(b)\mathcal{M} \vdash \mathcal{N}, \chi \Rightarrow \mathcal{M} \vdash \mathcal{N}, \phi \vee \chi$.

DJ-L: $\phi, \mathcal{M} \vdash \mathcal{N}; \chi, \mathcal{M} \vdash \mathcal{N} \Rightarrow \phi \vee \chi, \mathcal{M} \vdash \mathcal{N}$.

CJ-R: $\mathcal{M} \vdash \mathcal{N}, \phi; \mathcal{M} \vdash \mathcal{N}, \chi \Rightarrow \mathcal{M} \vdash \mathcal{N}, \phi \wedge \chi$.

CJ-L: $(a)\phi, \mathcal{N} \vdash \mathcal{N} \Rightarrow \phi \wedge \chi, \mathcal{M} \vdash \mathcal{N}$ - $(b)\chi, \mathcal{M} \vdash \mathcal{N} \Rightarrow \phi \wedge \chi, \mathcal{M} \vdash \mathcal{N}$.

CD-R: $\phi, \mathcal{M} \vdash \mathcal{N}, \chi \Rightarrow \mathcal{M} \vdash \mathcal{N}, \phi \rightarrow \chi$.

CD-L: $\mathcal{M} \vdash \mathcal{N}, \phi; \chi, \mathcal{M} \vdash \mathcal{N} \Rightarrow \phi \rightarrow \chi, \mathcal{M} \vdash \mathcal{N}$.

ULQ-R: $X \in W, \mathcal{M} \vdash \mathcal{N}, \phi(Y/X) \Rightarrow \mathcal{M} \vdash \mathcal{N}, (\forall Y \in W)\phi(Y)$,

where the substitution is free, the variable Y does not occur in W, and the variable X does not occur free in any of the components of $\mathcal{M}$, $\mathcal{N}$,$(\forall Y \in W)\phi(Y)$.

ULQ-L: $\mathcal{M} \vdash \mathcal{N}, V \in W; \phi(Y/V), \mathcal{M} \vdash \mathcal{N} \Rightarrow (\forall Y \in W)\phi(Y), \mathcal{M} \vdash \mathcal{N}$,

where the substitution is free, and the variable Y does not occur in W.

ELQ-R: $\mathcal{M} \vdash \mathcal{N}, V \in W; \mathcal{M} \vdash \mathcal{N}, \phi(Y/V) \Rightarrow \mathcal{M} \vdash \mathcal{N}, (\exists Y \in W)\phi(Y)$,

where the substitution is free, and the variable Y does not occur in W.

ELQ-L: $X \in W, \phi(Y/X), \mathcal{M} \vdash \mathcal{N} \Rightarrow (\exists Y \in W)\phi(Y), \mathcal{M} \vdash \mathcal{N}$,

where the substitution is free, the variable Y does not occur in W, and the variable X does not occur free in any of the components of $\mathcal{M}$, $\mathcal{N}$, $(\exists Y \in W)\phi(Y)$.

UGQ-R: $\mathcal{M} \vdash \mathcal{N}, \phi(Y/X) \Rightarrow \mathcal{M} \vdash \mathcal{N}, (\forall Y)\phi(Y)$,

where the substitution is free, and the variable X does not occur free in any component of $\mathcal{M}$, $\mathcal{N}$, $(\forall Y)\phi(Y)$.

UGQ-L: $\phi(Y/V), \mathcal{M} \vdash \mathcal{N} \Rightarrow (\forall Y)\phi(Y), \mathcal{M} \vdash \mathcal{N}$,

where the substitution is free.

EGQ-R: $\mathcal{M} \vdash \mathcal{N}, \phi(Y/V) \Rightarrow \mathcal{M} \vdash \mathcal{N}, (\exists Y)\phi(Y)$,

where the substitution is free.

EGQ-L: $\phi(Y/X), \mathcal{M} \vdash \mathcal{N} \Rightarrow (\exists Y)\phi(Y), \mathcal{M} \vdash \mathcal{N}$,

where the substitution is free and the variable X does not occur free in any of the components of $\mathcal{M}$, $\mathcal{N}$,$(\exists Y)\phi(Y)$.

The above rules define the *classical calculus of sequents.* A sequent that can be obtained by a finite number of applications of the rules, starting with initial sequents, is said to be *derivable in the classical calculus.* The derivation can be organized as a tree, where the root at the bottom is the derived sequent, and the leaves at the top are initial sequents. The length of the longest branch in the tree, not counting sequents obtained by applications of rule **INC**, is called the *height* of the derivation. The height is a measure of the complexity of the derivation.

The rule **INC** is called the *inclusion rule.* A rule different from the inclusion rule is said to be a *proper rule.* Every proper rule is either a right rule (indicated by the letter **R**), or a left rule (indicated by the letter **L**). Rule **INC** plays a crucial role in the calculus, although it is ignored in the evaluation of the height. In particular, the rule allows for the identification of several instances of the same formula generated in the derivation.

The quantifier rules have been written in such a way that the variable Y can be taken arbitrarily. In particular, the variables Y and X in some of these rules can be identified. The idea is that a formula $\phi(Y/V)$ (or $\phi(Y/X)$) is identical with a formula $\phi(Y/Z)(Z/V)$ (or $\phi(Y/Z)(Z/X)$) if Z is a new variable. This immediately induces a rule of change of bound variables, which is a useful tool in the theory of Gentzen systems. This rule is not used in this presentation, which relies mostly on the change of free variables (see Remarks 9.1 and 9.2).

EXERCISE 9.2 Prove that if ϕ is a formula, then $\phi \vdash \phi$ is derivable in the classical calculus.

EXAMPLE 9.1 Consider the following derivation in the classical calculus, where ϕ is an arbitrary formula:

$\phi \vdash \phi$ (from Exercise 9.2)

$\phi \vdash \phi \vee \neg\phi$ (rule **DJ-R** (a))

$\vdash \phi \vee \neg\phi, \neg\phi$ (rule **NG-R**)

$\vdash \phi \vee \neg\phi, \phi \vee \neg\phi$ (rule **DJ-R**(b))

$\vdash \phi \vee \neg\phi$ (rule **INC**)

Note that the end sequent is deterministic, but several intermediate sequents are non-deterministic. □

EXERCISE 9.3 Let $\phi(Y/X)$ be the formula occurring in rule **ULQ-R**. Prove that $\phi(Y/X)(X/Y)$ is the formula $\phi(Y)$. Prove the same result for rules **ELQ-L**, **UGQ-R**, and **EGQ-L**.

EXERCISE 9.4 Prove that if a variable occurs bound in some sequent of a derivation, then it occurs bound at the bottom of the derivation. Is this result true for variables that occur free in some sequent of the derivation?

EXERCISE 9.5 Assume a sequent $\mathcal{M} \vdash \phi$ is derivable in the classical calculus, and the variable Y does not occur free in $\mathcal{M}$. Prove that $\mathcal{M} \vdash (\forall Y)\phi$ is also derivable. Prove also that if W is a term that does not contain the variable Y, then $\mathcal{M} \vdash (\forall Y \in W)\phi$ is also derivable.

EXERCISE 9.6 Consider an application of rule **UGQ-R** of the form

$$\mathcal{M} \vdash \mathcal{N}, \phi(Y/X) \Rightarrow \mathcal{M} \vdash \mathcal{N}, (\forall Y)\phi(Y)).$$

Assume Z is a new variable and prove that the following is also a valid application of rule **UGQ-R**:

$$\mathcal{M} \vdash \mathcal{N}, \phi(Y/X) \Rightarrow \mathcal{M} \vdash \mathcal{N}, (\forall Z)\phi(Y/Z).$$

9.3 Derivations

In this section we study derivations in the calculus of sequents and describe several transformations where from given derivations we obtain new derivations with the same height that satisfy special requirements. These results apply to the classical calculus and also to the intuitionistic calculus defined in the next chapter.

In several of the quantifier rules there are restrictions on one of the variables, which is always denoted by the symbol X and does not appear free in the conclusion of the rule (rules **ULQ-R**, **ELQ-L**, **UGQ-R**, and **EGQ-L**), although it may appear free again in a later sequent of the derivation. Such a variable is called the *characteristic variable of the rule.* For example, in rule **EGQ-L** the formula $\phi(Y/X)$ induces the new formula $(\exists Y)\phi(Y)$, and the characteristic variable X disappears, as a free variable, from the new sequent generated by the rule. As explained above, this situation can be used to change the bound variable Y to another new variable Z, but it can also be used to change the characteristic variable X to another characteristic variable Z. In fact, the system offers plenty of freedom to change free variables within reasonable restrictions.

Remark 9.1 Given a derivation $\mathcal{D}$ for a sequent $\mathcal{M} \vdash \mathcal{N}$, it may happen that a variable X occurs free in some sequent of $\mathcal{D}$ but does not occur free in $\mathcal{M} \vdash \mathcal{N}$. If this is the case we can take a new variable Z that does not occur at all in the derivation (free or bound) and replace all free occurrences of X with Z. It is easy to verify that all rules are preserved, so we have a new derivation of the same height for the sequent $\mathcal{M} \vdash \mathcal{N}$. In this transformation it is possible, but not necessary, that the variable X be a characteristic variable in some rule of the derivation. The only condition is that the variable X does not occur free in the sequent $\mathcal{M} \vdash \mathcal{N}$. □

Remark 9.2 A similar transformation can be performed if X is the characteristic variable in some rule of the derivation for a sequent $\mathcal{M} \vdash \mathcal{N}$, even if X occurs free in $\mathcal{M} \vdash \mathcal{N}$. In fact, we can change X to another characteristic variable Z and we need only to perform the replacement of X by Z on the partial derivation above the rule, leaving the other occurrences untouched. Again, this replacement will not affect the height of the derivation. □

Theorem 9.1 *If there is a derivation of height k for a sequent $\mathcal{M} \vdash \mathcal{N}$, and no variable in the term V is bound in any of the components of $\mathcal{M}$ and $\mathcal{N}$, then there is a derivation of height k for the sequent $\mathcal{M}(X/V) \vdash \mathcal{N}(X/V)$.*

PROOF. From the assumptions we know that no variable in the term V occurs bound in the whole derivation. Furthermore, as explained in Remark 9.2, we can change the derivation, without affecting the height k, and make sure that neither the variable X nor the variables in V are characteristic variables. Now it is easy to verify that if we replace every component χ in the derivation by $\chi(X/V)$, all the rules are preserved and we get a derivation for $\mathcal{M}(X/V) \vdash \mathcal{N}(X/V)$. □

Theorem 9.2 *Assume ϕ is a formula of the form $(\forall Y)\phi'(Y)$, there is a derivation of height k for a sequent $\mathcal{M} \vdash \mathcal{N}$, and the variable Z does not occur free or bound in the derivation. If $\mathcal{N}'$ is the result of deleting all occurrences of ϕ in $\mathcal{N}$, then there is a derivation of height $k' \leq k$ for the sequent $\mathcal{M} \vdash \phi'(Y/Z), \mathcal{N}'$.*

PROOF. We obtain the new derivation from the given derivation of height k for the sequent $\mathcal{M} \vdash \mathcal{N}$. First, we make sure that no variable free in the formula $\phi'(Y/Z)$ is a characteristic variable in the given derivation. Next, we proceed to replace every sequent $\mathcal{M}_1 \vdash \mathcal{N}_1$ in the derivation by the sequent $\mathcal{M}_1 \vdash \phi'(Y/Z), \mathcal{N}_1'$ where $\mathcal{N}_1'$ is the result of deleting all occurrences of ϕ in

$\mathcal{N}_1$. The result is not a derivation, for it is possible to have rules involving ϕ that are affected by the change. This may happen in two ways. If ϕ is a component in the rule (for example, to introduce $\neg\phi$ in the left by rule **NG-L**), we reintroduce ϕ using the inclusion rule. The other possibility is that ϕ is introduced by rule **UGQ-R**. We consider the first such occurrence from top to bottom of the derivation, which originally is of the form:

$$\mathcal{M}_1 \vdash \mathcal{N}_1, \phi'(Y/X) \Rightarrow \mathcal{M}_1 \vdash \mathcal{N}_1, (\forall Y)\phi'(Y),$$

but now is of the form:

$$\mathcal{M}_1 \vdash \phi'(Y/Z), \mathcal{N}_1', \phi'(Y/X) \Rightarrow \mathcal{M}_1 \vdash \phi'(Y/Z), \mathcal{N}_1'.$$

We replace all occurrences of the characteristic variable X by the variable Z, but only on the partial derivation ending in this rule. The result takes the form

$$\mathcal{M}_1 \vdash \phi'(Y/Z), \mathcal{N}_1', \phi'(Y/Z) \Rightarrow \mathcal{M}_1 \vdash \phi'(Y/Z), \mathcal{N}_1',$$

which is obviously an application of the inclusion rule. Note that after this transformation the variable Z is not a characteristic variable in the derivation.

We continue this process with all the rules that introduce the formula ϕ. The result is a derivation of the sequent $\mathcal{M} \vdash \phi'(Y/Z), \mathcal{N}'$. □

Corollary 9.2.1 *Assume ϕ is a formula of the form $(\forall Y_1)...(\forall Y_s)\phi'(Y_s)...(Y_1)$, $1 < s$, there is a derivation of height k for a sequent $\mathcal{M} \vdash \mathcal{N}$, and the variables $Z_1, ..., Z_s$ do not occur free or bound in the derivation. If $\mathcal{N}'$ is the result of deleting all the occurrences of ϕ in $\mathcal{N}$, then there is a derivation of height $k' \leq k$ for the sequent $\mathcal{M} \vdash \phi'(Y_s/Z_s)...(Y_1/Z_1), \mathcal{N}'$.*

PROOF. Apply s times the transformation in Theorem 9.2. □

EXERCISE 9.7 Prove a result similar to Theorem 9.2 for the case ϕ is of one of the following forms: $(\exists Y)\phi'(Y)$, $(\forall Y \in W)\phi'(Y)$, and $(\exists Y \in W)\phi'(Y)$.

EXERCISE 9.8 Assume ϕ is a formula of the form $\psi \to \chi$ and there is a derivation of height k for a sequent $\mathcal{M} \vdash \mathcal{N}$. Prove that there is a derivation of height $k' \leq k$ for the sequent $\mathcal{M}, \psi \vdash \chi, \mathcal{N}'$, where $\mathcal{N}'$ is obtained from $\mathcal{N}$ by deleting all occurrences of ϕ.

If ϕ is a formula and there is a derivation in the classical calculus of a sequent $\mathcal{M} \vdash \phi$ where all the components of $\mathcal{M}$ are closed axioms, we say that ϕ is a *formal theorem in classical logic.* It can be proved that the axioms derived by rule **ST 15** are not necessary, but we do not need this result here.

There are several important properties of the classical calculus that are not proved here, but are available in the literature.

In the first place, the classical calculus is closed under the cut rule. This means that if sequents $\mathcal{M}_1 \vdash \mathcal{N}_1, \phi$ and $\phi, \mathcal{M}_2 \vdash \mathcal{N}_2$ are derivable, where no variable occurs both free and bound, then the sequent $\mathcal{M}_1, \mathcal{M}_2 \vdash \mathcal{N}_1, \mathcal{N}_2$ is also derivable.

The calculus is semantically complete, relative to the usual first order semantics. This is not immediately relevant to our investigation, since we reject the notion of a complete universe of sets. Still, it means that all theorems proved in this work can be formulated as formal theorems in classical logic.

EXAMPLE 9.2 In Chapter 3 we proved in Theorem 3.3 a formula ϕ of the form $(\forall Z)\phi' \to (\forall Z)\phi''$, where ϕ' and ϕ'' are local formulas. By completeness, this means that there is a classical derivation for a sequent $\mathcal{M} \vdash \phi$ where the components of $\mathcal{M}$ are closed axioms. □

9.4 Local Reduction

We know that a local formula does not contain global quantifiers. A formula ϕ of the form $(\forall Y_1)...(\forall Y_s)\phi'$, $0 \leq s$, where ϕ' is local, is said to be a Π-*formula.* If ϕ is of the form $(\exists Y_1)...(\exists Y_s)\phi'$ where ϕ' is local, we say that

ϕ is a *Σ-formula*. A local formula is both a Π-formula and a Σ-formula. Note that if ϕ is a Π-formula (Σ-formula), then every sub-formula of ϕ is a Π-formula (Σ-formula). Clearly, all the closed axioms defined in Section 9.1 are Π-formulas.

A sequent $\mathcal{M} \vdash \mathcal{N}$ where all the components of $\mathcal{M}$ are Π-formulas and all the components of $\mathcal{N}$ are Σ-formulas is said to be a *Δ-sequent*. In a derivation for a Δ-sequent all the sequents are Δ-sequents. From this it follows that there are rules that cannot occur in a derivation for a Δ-sequent, namely, the rules **UGQ-R** and **EGQ-L**.

Let ϕ be an existential formula of the form $(\exists Y)\phi'$, where ϕ' is an arbitrary formula. We define the *local reductions of ϕ* by the following rules:

LR 1 If V is a term, then $\phi'(Y/V)$ is a local reduction of ϕ, provided the substitution is free.

LR 2 If χ_1 and χ_2 are local reductions of ϕ, then $\chi_1 \vee \chi_2$ is a local reduction of ϕ.

LR 3 If $\chi(Y'/X)$ is a local reduction of ϕ, the substitution is free, and W is a term, then $(\exists Y' \in W)\chi(Y')$ is a local reduction of ϕ, provided the variable Y' does not occur in W and the variable X does not occur free in ϕ or $(\exists Y' \in W)\chi(Y')$.

Note that if ϕ' is local, then all the local reductions of ϕ are local.

Remark 9.3 If χ is a local reduction of ϕ, then every free variable of ϕ occurs free in χ, but χ may contain free variables that do not occur free in ϕ. The new variables in χ enter via the term V in rule **LR 1**, and the term W in rule **LR 3**. The characteristic variable X, eliminated by rule **LR 3**, is one of the new variables, and for this reason the construction of χ preserves all the free variables in ϕ. □

Theorem 9.3 *If χ is a local reduction of ϕ, then the sequent $\chi \vdash \phi$ is derivable in the classical calculus.*

PROOF. The proof is by induction on the construction of χ. In rule **LR 1** we use derivation rule **EGQ-R**. If χ is $\chi_1 \vee \chi_2$ where $\chi_1 \vdash \phi$ and $\chi_2 \vdash \phi$ are derivable, we use rule **DJ-L**. Finally, if χ is $\chi(Y'/X)$ in rule **LR 3**, and we know $\chi(Y'/X) \vdash \phi$ is derivable, then $(\exists Y' \in W)\chi(Y') \vdash \phi$ follows by rule **ELQ-L**, noting that the variable X does not occur free in ϕ. □

Remark 9.4 As stated, Theorem 9.3 involves only the classical calculus. In fact, the same proof is valid in the intuitionistic calculus, to be defined in the next chapter. □

Theorem 9.4 *Assume there is a derivation in the classical calculus for a Δ-sequent $\mathcal{M} \vdash \mathcal{N}$ and ϕ is a non-local Σ-formula. There is a derivation in the classical calculus for a sequent $\mathcal{M} \vdash \mathcal{N}'$, where the sequence $\mathcal{N}'$ is derived from $\mathcal{N}$ by deleting all occurrences of ϕ and adding some local reductions of ϕ.*

PROOF. The proof is by induction on the height k of the derivation for $\mathcal{M} \vdash \mathcal{N}$. We assume ϕ is of the form $(\exists Y)\phi'$. If ϕ does not occur in $\mathcal{N}$ we take $\mathcal{N}'$ equal to $\mathcal{N}$. So, we assume ϕ is a component of $\mathcal{N}$. The proof consists of a number of cases, each determined by the rule that introduces the sequent $\mathcal{M} \vdash \mathcal{N}$. In general, they are trivial and require only to apply the induction hypothesis to the preceding sequents and then apply the same rule again. Rules **UGQ-R** and **EGQ-L** are impossible and there are only three rules that require special argument, one of which is **EGQ-R** (when this rules introduces ϕ). The others are **ULQ-R** and **ELQ-L**, because they involve characteristic variables that can be affected by the introduction of new local reductions of ϕ.

The case of rule **EGQ-R**, when the formula ϕ is introduced by the rule, takes the form,

$$\mathcal{M} \vdash \mathcal{N}_1, \phi'(Y/V) \Rightarrow \mathcal{M} \vdash \mathcal{N}_1, \phi.$$

Note that $\phi'(Y/V)$ is a local reduction of ϕ, and by the induction hypothesis we have a derivation for $\mathcal{M} \vdash \mathcal{N}_1', \phi'(Y/V)$. So, we take $\mathcal{N}' = \mathcal{N}_1', \phi'(Y/V)$.

In the case of rule **ULQ-R** the sequent $\mathcal{M} \vdash \mathcal{N}$ is derived by the following application of the rule,

$$X \in W, \mathcal{M} \vdash \mathcal{N}_1, \psi(Y'/X), \Rightarrow \mathcal{M} \vdash \mathcal{N}_1, (\forall Y' \in W)\psi(Y'),$$

where $\mathcal{N}_1, (\forall Y' \in W)\psi(Y')$ is the sequence $\mathcal{N}$, and we assume ϕ is a component of $\mathcal{N}_1$. By the induction hypothesis we have a derivation for the following sequent, where $\mathcal{N}_1'$ is obtained from $\mathcal{N}_1$ by deleting all occurrences of ϕ and adding local reductions of ϕ,

$$X \in W, \mathcal{M} \vdash \mathcal{N}_1', \psi(Y'/X).$$

We would like to apply rule **ULQ-R** again, but this may be impossible if there is a partial reduction χ that contains free occurrences of the variable X. If this is the case, we write $\mathcal{N}_1' = \mathcal{N}_1'', \chi$ and write the following derivation:

(1) $X \in W \vdash X \in W$ (initial sequent).

(2) $X \in W, \mathcal{M} \vdash \mathcal{N}_1'', \psi(Y'/X), \chi(X)$.

(3) $X \in W, \mathcal{M} \vdash \mathcal{N}_1'', \psi(Y'/X), (\exists X \in W)\chi(X)$ (rule ELQ-R).

Note that $(\exists X \in W)\chi(X)$ is also a partial reduction of ϕ, because the variable X does not occur free in ϕ, which we assume to be a component of the sequent. By this procedure we eliminate all the free occurrences of the variable X in partial reductions of ϕ. To complete the proof of the case we again apply rule **ULQ-R**.

The case of rule **ELQ-L** is similar to the preceding case.

Corollary 9.4.1 *Assume ϕ is a formula of the form $(\exists Y)\phi'$, where ϕ' is local, and ϕ is a formal theorem in classical logic. There is a local formula χ such that χ is a formal theorem in classical logic, and $\chi \vdash \phi$ is derivable in the classical calculus*

PROOF. From Theorem 9.4 we know that there is a derivation of $\mathcal{M} \vdash \chi$, where $\chi = \chi_1 \vee \ldots \vee \chi_s$ is a local reduction of ϕ. From Theorem 9.3 it follows that $\chi \vdash \phi$ is derivable in the classical calculus. □

EXERCISE 9.9 Prove that if χ is a local reduction of ϕ, the variable X does not occur free in ϕ, and Z is a variable, then $\chi(X/Z)$ is a local reduction of ϕ, provided the substitution is free.

Theorem 9.5 *Let ϕ be an existential formula of the form $(\exists Y)\phi'$, and χ a local reduction of ϕ. If Z is a variable that does not occur free in χ, then there is a local reduction χ' of the form $(\exists Z \in W)\phi'(Y/Z)$ and $\chi \to \chi'$ is a formal theorem in classical logic.*

PROOF. The proof is by induction on the complexity of χ. Here we apply informally the axioms of the system $\mathbb{G}$ and leave to the reader the trivial formalization by means of the formal axioms in Section 9.1. If χ is of the form $\phi'(Y/V)$ we take χ' the formula $(\exists Z \in \{V\})\phi'(Y/Z)$, and this involves the axioms for the empty set and the insertion operation. If χ is of the form $\chi_1 \vee \chi_2$, and we have defined χ_1' of the form $(\exists Z \in W_1)\phi'(Y/Z)$ and χ_2' of the form $(\exists Z \in W_2)\phi'(Y/Z)$, we take χ' the formula $(\exists Z \in W_1 \cup W_2)\phi'(Y/Z)$, and this involves empty set, insertion, and replacement. Finally, if χ is of the form $(\exists Y' \in W_1)\chi_1(Y')$, derived by rule **LR 3** from $\chi_1(Y'/X)$, we may assume that the variable Z is different from X, hence by the induction hypothesis there is local reduction of the form $(\exists Z \in W_2)\phi'(Y/Z)$, and $(\exists Z \in W_2)\phi'(Y/Z)$ is a formal theorem in classical logic. We conclude that $(\exists Y' \in W_1)\chi_1(Y') \to (\exists X \in W_1)(\exists Z \in W_2)\phi'(Y/Z)$ is also a formal theorem in intuitionistic logic. Using the axioms of the theory, we can show that the following is a formal theorem in classical logic:

$$\chi \to (\exists Z \in \bigcup\{W_2 : X \in W_1 :\})\phi'(Y/Z),$$

and this involves replacement. Hence, in this case W is the term $\bigcup\{W_2 : X \in W_1 :\}$. □

Remark 9.5 The construction in Theorem 9.5 is independent of the formula ϕ', which is completely arbitrary. The term W is determined by the terms entering the construction of χ. To determine χ we need Theorem 9.4, but this requires a formal proof of ϕ, so ϕ' must be a local formula. Still, it is a practical result. For example, assume p is a binary predicate in the system $\mathbb{G}$ (so $p(X,Y)$ is essentially a local formula), and we have a formal proof of $\psi \rightarrow (\exists Y)p(X,Y)$, where ψ is a Π-formula that contains the free variable X. Since the classical calculus is complete, there is a derivation in the classical calculus of a sequent $\mathcal{M}, \psi \vdash (\exists Y)p(X,Y)$. Using Theorems 9.3, 9.4 and 9.5 we conclude that there is a term W such that $\psi \rightarrow (\exists Y \in W)p(X,Y)$ is a formal theorem in classical logic, so it is a theorem in the system $\mathbb{G}$. If we set $g(X) = W$, we can introduce the operation f such that,

$$f(X) = \bigcup\{Y \in g(X) :: p(X,Y)\}.$$

Now, if X is set such that ψ holds and there is a unique Y such that $p(X,Y)$ holds, it follows that $f(X) = Y$, hence $p(X, f(X))$ holds for that particular X. The operation f is not an addition to the system $\mathbb{G}$. Rather, the operation f is a basic operation derived from the basic rules in the system $\mathbb{G}$. We can see that the derivation of $\mathcal{M}, \psi \vdash (\exists Y \in W)p(X,Y)$ provides the machinery necessary to identify f. □

9.5 Notes

The formalization of the underlying logic is a crucial element in the system we propose here, for we consider set theory to be a foundational system, and its logic must be an integral element to be defined with the axioms. In other words, we cannot assume a logic existing before set theory, for such a logic needs a semantics which can be given only in the context of some set theory. This means that the underlying logic must be a part of the system $\mathbb{G}$ and given by some formal system. In general, we propose that such a system must be the intuitionistic system described in the next chapter. Still, we are interested in the use of classical logic to prove local theorems, and we show in the next chapter that such theorems can be proved in a partial system of intuitionistic logic, where we allow for the axioms derived from rule **ST**

15. Note that the only non-local theorem proved in the text is Theorem 3.3, discussed in Example 9.2.

Both classical logic and intuitionistic logic are formalized via the well-known calculus of sequents, introduced in Gentzen [9]. We assume some familiarity with this calculus, the most important reference being Curry (1963), who provides an exhaustive analysis of each connective and quantifier. Kleene (1952) gives a complete proof of the so-called elimination theorem for the classical and for the intuitionistic calculus. In the next chapter we give a similar proof for the intuitionistic calculus, for this result plays a crucial role in the reduction theorem. All these proofs are derived from the original proof given in Gentzen [9]

Theorem 9.4 is a local reduction theorem for the classical calculus, and it is extremely useful in the system $\mathbb{G}$, where we require local quantification in the replacement rule, proofs by induction, and the completeness axiom for the ordinal permutation rule. In Remark 9.5 we show how to use this result to derive set operations with selector properties relative to some (local) predicate p.

Chapter 10

Formalization: Intuitionistic Logic

We have explained before that classical logic is not the right tool to deal with global formulas (see Remark 1.1). A similar position is taken in intuitionistic mathematics, although in a more general setting. In fact, in intuitionistic mathematics there is a general objection to using classical logic in dealing with infinite sets. We do not share this radical position, and we think that classical logic is admissible as long as we are dealing with well-defined totalities, finite or infinite. On the other hand, we think the universe of sets is not a well-defined totality, for it is not supported by an objective construction. Still, the system $\mathbb{G}$ allows for references to the universe, and in fact the crucial Theorem 3.3 is of the form $(\forall Z)\phi' \rightarrow (\forall Z)\phi''$, where ϕ' and ϕ'' are local formulas (see Example 9.2). To account for the peculiar role of the universe, we propose that in a general context that involves the universe we must use the principles of intuitionistic logic, while in a local context we can use classical logic.

It is a well-known fact that intuitionistc logic can be described as a subset of classical logic that is formally determined by restricting the calculus to deterministic sequents. We pursue this approach here and prove one important classical reduction theorem. Furthermore, we extend the local reduction theorem of Chapter 9.

10.1 The Intuitionistic Calculus

We obtain the intuitionistic calculus by restricting the rules of the classical calculus to deterministic sequents. The initial sequents are clearly deterministic. In general, the derivation rules of the classical calculus may need some restrictions to make sure that they produce a deterministic sequent whenever applied to deterministic sequents. For example, in rule **INC** we must require that $\mathcal{N}'$ be deterministic, in rule **NG-R** that $\mathcal{N}$ be empty, etc. In any case, the intuitionistic calculus is a sub-system of the classical calculus, and the notation introduced for the classical calculus (characteristic variable, height, etc.) applies also to the intuitionistic calculus.

EXERCISE 10.1 Let ϕ be an arbitrary formula. Prove that $\phi \vdash \phi$ is derivable in the intuitionistic calculus.

EXERCISE 10.2 Let $\phi(X)$ be a an arbitrary formula. Prove that the sequent $(\forall X)\phi(X) \vdash \neg(\exists X)\neg\phi(X)$ is derivable in the intuitionistic calculus.

EXAMPLE 10.1 Consider the following simplified derivation in the intuitionistic calculus, where ϕ_1 and ϕ_2 are arbitrary formulas:

(1) $\phi_1 \vdash \phi_1$ (holds for arbitrary formula).

(2) $\neg\phi_1, \phi_1 \vdash$ (rule **NG-L**).

(3) $\phi_1, \neg\phi_1 \vdash \phi_2$ (rule **INC**).

(4) $\phi_2 \vdash \phi_2$.

(5) $\phi_1 \vee \phi_2, \neg\phi_1 \vdash \phi_2$ (rule **DJ-L** from (3) and (4)).

This shows that the sequent $\phi_1 \vee \phi_2, \neg\phi_1 \vdash \phi_2$ is derivable in the intuitionistic calculus. □

EXERCISE 10.3 Assume $\phi(X)$ is an arbitrary formula. Prove that the sequent $\neg\neg(\forall X)\phi(X) \vdash (\forall X)\neg\neg\phi(X)$ is derivable in the intuitionistic calculus.

EXERCISE 10.4 Let $\phi(X)$ be an arbitrary formula. Prove that the sequent $\neg(\exists X)\neg\phi(X) \vdash (\forall X)\neg\neg\phi(X)$ is derivable in the intuitionistic calculus.

EXERCISE 10.5 Assume ϕ is a formula of the form $(\exists Y)\phi'(Y)$, there is a derivation of height k in the intuitionistic calculus for a sequent $\mathcal{M} \vdash \mathcal{N}$, and the variable Z does not occur free or bound in the derivation. Prove that there is a derivation in the intuitionistic calculus of height $k' \leq k$ for the sequent $\mathcal{M}', \phi'(Y/Z) \vdash \mathcal{N}$, where $\mathcal{M}'$ is the result of deleting all occurrences of ϕ in $\mathcal{M}$.

If ϕ is a formula and there is a derivation in the intuitionistic calculus of a sequent $\mathcal{M} \vdash \phi$ where all the components of $\mathcal{M}$ are closed axioms, we say that ϕ is a *formal theorem in intuitionistic logic.* We know that all results proved in this monograph are formal theorems in classical logic. We show later that all are formal theorems in intuitionistic logic.

EXERCISE 10.6 Let p be k-ary predicate constant. Prove that $p(\mathbf{X}) \vee \neg p(\mathbf{X})$ is a formal theorem in intuitionistic logic.

EXERCISE 10.7 Assume the sequent $\phi(Y/X) \vdash \psi(Z/X)$ is derivable in the intuitionistic calculus, where the variable X does not occur free or bound in ϕ or ψ. Prove that if W is a term, and neither Y nor Z occurs in W, then the sequent $(\exists Y \in W)\phi(Y) \vdash (\exists Z \in W)\psi(Z)$ is also derivable.

Theorem 10.1 *If there is a derivation of height k in the intuitionistic calculus for a sequent $\mathcal{M} \vdash \mathcal{N}$, and no variable in the term V is bound in any of the components of $\mathcal{M}$ and $\mathcal{N}$, then there is a derivation of height k for the sequent $\mathcal{M}(X/V) \vdash \mathcal{N}(X/V)$.*

PROOF. Same proof as for Theorem 9.1. □

We say that a finite set $\mathcal{S}$ of sequents is *compatible* if there is no variable that occurs free and bound in components of elements of $\mathcal{S}$. If $\mathcal{S} = \{S\}$ is a singleton, we say that S is compatible. If $\mathcal{S} = \{S_1, ..., S_n\}$ we say that $S_1, ..., S_n$ are compatible. A derivation $\mathcal{D}$ is compatible if the set of all sequents in the derivation is compatible. Two derivations, $\mathcal{D}_1$ and $\mathcal{D}_2$, are compatible if the set of all sequents in the derivations is compatible.

Theorem 10.2 *Assume there are derivations for the sequents S_1 and S_2, where S_1 and S_2 are compatible. Then there are derivations of the same height that are also compatible.*

PROOF. Using the procedure described in Remarks 9.1 and 9.2 we can change the free variables in the given derivations until compatible derivations are obtained. □

We say that a formula ϕ has the *cut property* in the intuitionistic calculus if whenever $\mathcal{M}_1 \vdash \phi$ and $\mathcal{M}_2 \vdash \mathcal{N}$ are compatible derivable sequents, then $\mathcal{M}_1, \mathcal{M}_2' \vdash \mathcal{N}$ is also derivable, where $\mathcal{M}_2'$ is the result of deleting all components ϕ from the sequence $\mathcal{M}_2$.

EXERCISE 10.8 Let ϕ be an atomic formula. Prove that ϕ has the cut property.

Theorem 10.3 *Every formula ϕ has the cut property.*

PROOF. The proof is by induction on the number of logical symbols occurring in the formula ϕ: $\neg, \vee, \wedge, \rightarrow, (\forall...), (\exists...), (\forall... \in ...)(\exists... \in ...)$. We fix ϕ and assume that whenever ψ is a formula with fewer logical symbols than ϕ, then ψ has the cut property. Under this assumption we prove the following assertion: if $\mathcal{M}_1 \vdash \phi$ has a derivation of height k_1, and $\mathcal{M}_2 \vdash \mathcal{N}$ has a derivation of height k_2, and the two sequents are compatible, then the sequent $\mathcal{M}_1, \mathcal{M}_2' \vdash \mathcal{N}$ is derivable, where $\mathcal{M}_2'$ is the result of deleting all occurrences of ϕ in the sequence $\mathcal{M}_2$.

We prove the assertion by induction on k_1+k_2. We note first that the case $k_1 = 1$ and the case $k_2 = 1$ are trivial. So we assume in general that $k_1 > 1$ and $k_2 > 1$. Now, with fixed k_1 and k_2 we take two sequents $S_1 = \mathcal{M}_1 \vdash \phi$ and $S_2 = \mathcal{M}_2 \vdash \mathcal{N}$ that are compatible and have derivations of height k_1 and k_2, respectively, that we assume are compatible.

(i) We shall assume that the last rule in the derivations of S_1 and S_2 is not the inclusion rule. The general case can be reduced to this by considering the partial derivations where the last rule is a proper rule. This is always possible, because $k_1 > 1$ and $k_2 > 1$.

(ii) S_1 is obtained by a unary left rule. Here we use the induction on k_1 and the same rule again. For example, consider the rule **ELQ-L**, so S_1 is the sequent $(\exists Y \in W)\psi, \mathcal{M}_1 \vdash \phi$ derived in the form,

$$X \in W, \psi(Y/X), \mathcal{M}_1 \vdash \phi \Rightarrow (\exists Y \in W)\psi(Y), \mathcal{M}_1 \vdash \phi.$$

We may assume that the characteristic variable X does not occur free in $\mathcal{M}_2, \mathcal{N}$. The following sequents are derivable:

(1) $X \in W, \psi(Y/X), \mathcal{M}_1, \mathcal{M}_2' \vdash \mathcal{N}$ (induction on k_1).

(2) $(\exists Y \in W)\psi(Y), \mathcal{M}_1, \mathcal{M}_2' \vdash \mathcal{N}$ (rule **ELQ-L**).

(iii) S_1 is obtained by a binary left rule. Again, the result follows by the induction on k_1 and the same rule. For example, consider the rule **CD-L**, where S_1 is the sequent $\psi_1 \to \psi_2, \mathcal{M}_1 \vdash \phi$ derived in the form,

$$\mathcal{M}_1 \vdash \psi_1; \psi_2, \mathcal{M}_1 \vdash \phi \Rightarrow \psi_1 \to \psi_2, \mathcal{M}_1 \vdash \phi.$$

The following sequents are derivable:

(1) $\psi_2, \mathcal{M}_1, \mathcal{M}_2' \vdash \mathcal{N}$ (induction on k_1).

(2) $\psi_1 \to \psi_2, \mathcal{M}_1, \mathcal{M}_2' \vdash \mathcal{N}$ (rules **INC** and **CD-L**).

(iv) S_2 is obtained by a right rule. For example, consider the case where S_2 is derived by rule **CD-R**, so S_2 is the sequent $\mathcal{M}_2 \vdash \chi_1 \to \chi_2$, obtained by the derivation,

$$\chi_1, \mathcal{M}_2 \vdash \chi_2 \Rightarrow \mathcal{M}_2 \vdash \chi_1 \to \chi_2.$$

The following sequents are derivable:

(1) $\mathcal{M}_1, \chi_1, \mathcal{M}_2' \vdash \chi_2$ (induction on k_2).

(2) $\mathcal{M}_1, \mathcal{M}_2' \vdash \chi_1 \to \chi_2$ (rules **INC** and **CD-R**).

We are ignoring the possibility that ϕ and χ_1 are the same formula, in which case the latter must be introduced again by inclusion.

Another example is the case where S_2 is obtained by rule **ELQ-R**, so S_2 is the sequent $\mathcal{M}_2 \vdash (\exists Y \in W)\chi(Y)$, obtained by the derivation,

$$\mathcal{M}_2 \vdash V \in W; \mathcal{M}_2 \vdash \chi(Y/V) \Rightarrow \mathcal{M}_2 \vdash (\exists Y \in W)\chi(Y).$$

The following sequents are derivable:

(1) $\mathcal{M}_1, \mathcal{M}_2' \vdash V \in W$ (induction on k_2).

(2) $\mathcal{M}_1, \mathcal{M}_2' \vdash \chi(Y/V)$ (induction on k_2).

(3) $\mathcal{M}_1, \mathcal{M}_2' \vdash (\exists Y \in W)\chi(Y)$ (rule **ELQ-R**).

(v) S_2 is obtained by a left rule that introduces a formula different from ϕ. These cases are handled by the induction on k_2 and are left to the reader.

(vi) S_1 is obtained by a right rule (so it introduces ϕ), and S_2 is obtained by a left rule that introduces ϕ. These cases require the induction on k_1 and k_2, and also the induction on the number of logical symbols in ϕ. There are exactly eight cases, one for each construction, and we discuss only a few of them. Note that at this stage the theorem has been proved for the case where ϕ is an atomic formula.

Consider the case where ϕ is the formula $\psi_1 \vee \psi_2$, so the derivation take the form,

(1) $\mathcal{M}_1 \vdash \psi_1 \Rightarrow \mathcal{M}_1 \vdash \psi_1 \vee \psi_2$.

(2) $\psi_1, \mathcal{M}_2 \vdash \mathcal{N}; \psi_2, \mathcal{M}_2 \vdash \mathcal{N} \Rightarrow \psi_1 \vee \psi_2, \mathcal{M}_2 \vdash \mathcal{N}$.

(3) $\mathcal{M}_1, \psi_1, \mathcal{M}_2' \vdash \mathcal{N}$ (induction on k_2).

(4) $\mathcal{M}_1, \mathcal{M}_2' \vdash \mathcal{N}$ (induction on ψ_1 and inclusion).

Note that since S_1 and S_2 are compatible, it follows that the sequents $\mathcal{M}_1 \vdash \psi_1$ and $\mathcal{M}_1, \psi_1, \mathcal{M}_2' \vdash \mathcal{N}$ are also compatible.

Assume now that ϕ is the formula $(\forall Y \in W)\psi(Y)$. We have the following derivations:

(1) $X \in W, \mathcal{M}_1 \vdash \psi(Y/X) \Rightarrow \mathcal{M}_1 \vdash (\forall Y \in W)\psi(Y)$.

(2) $\mathcal{M}_2 \vdash V \in W; \psi(Y/V), \mathcal{M}_2 \vdash \mathcal{N} \Rightarrow (\forall Y \in W)\psi(Y), \mathcal{M}_2 \vdash \mathcal{N}$.

(3) $V \in W, \mathcal{M}_1 \vdash \psi(Y/V)$ (Theorem 10.1).

(4) $\mathcal{M}_1, \mathcal{M}_2' \vdash V \in W$ (induction on k_2).

(5) $\mathcal{M}_1, \psi(Y/V), \mathcal{M}_2' \vdash \mathcal{N}$ (induction on k_2).

(6) $V \in W, \mathcal{M}_1, \mathcal{M}_2' \vdash \mathcal{N}$ (induction on $\psi(Y/V)$).

(7) $\mathcal{M}_1, \mathcal{M}_2' \vdash \mathcal{N}$ (induction on atomic $V \in W$).

We consider also the case where ϕ is the formula $(\exists Y \in W)\psi(Y)$, and we have the derivations;

(1) $\mathcal{M}_1 \vdash V \in W; \mathcal{M}_1 \vdash \psi(Y/V) \Rightarrow \mathcal{M}_1 \vdash (\exists Y \in W)\psi(Y)$.

(2) $X \in W, \psi(Y/X), \mathcal{M}_2 \vdash \mathcal{N} \Rightarrow (\exists Y \in W)\mathcal{M}_2 \vdash \mathcal{N}$.

(3) $V \in W, \psi(Y/V), \mathcal{M}_2 \vdash \mathcal{N}$ (Theorem 10.1).

(4) $\mathcal{M}_1, V \in W, \psi(Y/V), \mathcal{M}_2' \vdash \mathcal{N}$ (induction on k_2).

(5) $\mathcal{M}_1, \psi(Y/V), \mathcal{M}_2' \vdash \mathcal{N}$ (induction on atomic $V \in W$).

(6) $\mathcal{M}_1, \mathcal{M}'_2 \vdash \mathcal{N}$ (induction on $\psi(Y/V)$).

The other cases are similar. This completes the proof of the theorem. □

Corollary 10.3.1 *If If $\mathcal{M} \vdash \phi$ and $\mathcal{M}', \phi \vdash \psi$ are compatible sequents derivable in the intuitionistic calculus, then $\mathcal{M}, \mathcal{M}' \vdash \psi$ is also derivable in the intuitionistic calculus.*

PROOF. Immediate from Theorem 10.3 using ϕ as the cut formula. □

Remark 10.1 Applications of Corollary 10.3.1 will be referred to as applications of the *cut rule*, with ϕ as the *cut formula*. □

Corollary 10.3.2 *If $\mathcal{M} \vdash \neg\neg\phi$ is a compatible sequent derivable in the intuitionistic calculus, then $\phi \vee \neg\phi, \mathcal{M} \vdash \phi$ is also derivable in the intuitionistic calculus.*

PROOF. We know that $\neg\phi, \neg\neg\phi \vdash$ is derivable in the intuitionistic calculus. Using the cut rule we have $\mathcal{M}, \neg\phi \vdash$, hence $\mathcal{M}, \neg\phi \vdash \phi$. On the other hand, we have $\phi \vdash \phi$, hence using rule **DJ-L** we get $\phi \vee \neg\phi, \mathcal{M} \vdash \phi$ is derivable in the intuitionistic calculus. □

Corollary 10.3.3 *If $\mathcal{M} \vdash \phi \vee \psi$ is a compatible sequent derivable in the intuitionistic calculus, then $\mathcal{M}, \neg\phi \vdash \psi$ and $\mathcal{M}, \neg\psi \vdash \phi$ are also derivable in the intuitionistic calculus.*

PROOF. Clearly, $\psi \vdash \psi$, and $\phi, \neg\phi \vdash \psi$ are derivable. Using rule **DJ-L** we have $\phi \vee \psi, \neg\phi \vdash \psi$, and by the cut rule we have $\mathcal{M}, \neg\phi \vdash \psi$. The dual sequent is similar. □

EXERCISE 10.9 Assume a compatible sequent $\vdash \phi \to \chi$ is derivable in the intuitionistic calculus. Prove that $\phi \vdash \chi$ is also derivable.

EXERCISE 10.10 Assume a compatible sequent $\phi \to \chi \vdash \psi$ is derivable in the intuitionistic calculus. Prove that $\chi \vdash \psi$ and $\neg\phi \vdash \psi$ are also derivable.

EXERCISE 10.11 Assume ψ, ϕ_1, ϕ_2 are compatible formulas. Prove that the sequent $\psi \vee \phi_1, \psi \vee \phi_2 \vdash \psi \vee (\phi_1 \wedge \phi_2)$ is derivable in the intuitionistic calculus.

10.2 Classical Reduction

A formula ϕ is *regular* if whenever $\neg\phi'$ or $\phi' \to \phi''$ is a sub-formula of ϕ, then ϕ' is a Π-formula. Clearly, if ϕ is a Π-formula or a Σ-formula, then ϕ is regular. All the sub-formulas of a regular formula are regular. On the other hand, a formula ϕ may contain sub-formulas that are not Π-formulas, and still be regular. For example, if ψ is a Π-formula, then $(\exists Y)(\forall X)\psi$ is regular, but not a Π-formula. Furthermore, $\neg(\exists Y)(\forall X)\phi$ and $\neg\neg(\exists Y)(\forall X)\phi$ are not regular.

EXERCISE 10.12 Prove that if a formula ϕ is regular, then every sub-formula of ϕ is regular.

A sequent $\mathcal{M} \vdash \mathcal{N}$ is *regular* if all the components in $\mathcal{M}$ are regular and all the components in $\mathcal{N}$ are local.

Remark 10.2 A derivation in the intuitionistic calculus for a regular sequent $\mathcal{M} \vdash \mathcal{N}$ may contain sequents that are not regular. For example, the following is a valid derivation, where ϕ is a local formula:

$$(\forall X)\phi \vdash (\forall X)\phi \Rightarrow \neg(\forall X)\phi, (\forall X)\phi \vdash,$$

where the first sequent is not regular and the second is regular. □

EXAMPLE 10.2 In Chapter 3 we proved in Theorem 3.3 a formula ϕ of the form $(\forall Z)\phi' \to (\forall Z)\phi''$, where ϕ' and ϕ'' are local formulas (see Example 9.2). We know this means that there is a classical derivation for a sequent $\mathcal{M} \vdash \phi$ where the components of $\mathcal{M}$ are closed axioms. This sequent is not regular, because ϕ is not a local formula. On the other hand, we know that there is a classical derivation for $\mathcal{M}, (\forall Z)\phi' \vdash \phi''$, and this is a regular sequent. □

In this section we apply Theorem 10.3 to show that if a regular sequent is derivable in the classical calculus, then it is also derivable in the intuitionistic calculus. We describe this relation as a reduction from the classical calculus to the intuitionistic calculus.

To formalize the notion of reduction we need some notation to deal with disjunctions induced by sequences of formulas. If $\mathcal{N}$ is a sequence $\phi_1, ..., \phi_n$, $n \geq 0$, we want to give a practical description of the general disjunction $\phi_1 \vee ... \vee \phi_n$. If $\mathcal{N}$ is deterministic we set $\mathcal{N}^\vee = \mathcal{N}$. If $\mathcal{N} = \mathcal{M}, \phi$, where $\mathcal{M}$ is non-empty, we set $\mathcal{N}^\vee = \mathcal{M}^\vee \vee \phi$.

EXERCISE 10.13 Assume $\mathcal{M}$ and $\mathcal{N}$ are sequents with a common component. Prove that the sequent $\mathcal{M} \vdash \mathcal{N}^\vee$ is derivable in the intuitionistic calculus.

Theorem 10.4 *If there is a derivation in the classical calculus for a compatible regular sequent $\mathcal{M} \vdash \mathcal{N}$, then there is a derivation in the intuitionistic calculus for a sequent $\mathcal{M}', \mathcal{M} \vdash \mathcal{N}^\vee$, where the components of $\mathcal{M}'$ are closed axioms derived from rule* **ST 15**.

PROOF. Note that it is sufficient to prove that the formulas in $\mathcal{M}'$ are of the form $\phi \vee \neg\phi$ with ϕ local. The universal closure can be introduced using rule **UGQ-L**.

The proof is by induction on k = the height of the derivation of the sequent $\mathcal{M} \vdash \mathcal{N}$. The case $k = 1$ is trivial, so we assume $k > 1$. If the last rule of the derivation is an inclusion from $\mathcal{M}_1 \vdash \mathcal{N}_1$, and we have a derivation for $\mathcal{M}', \mathcal{M}_1 \vdash \mathcal{N}_1^\vee$ in the intuitionistic calculus, then either $\mathcal{N}_1^\vee$ is empty or $\mathcal{N}_1^\vee \vdash \mathcal{N}^\vee$ is also derivable, hence by cut and inclusion we get $\mathcal{M}', \mathcal{M} \vdash \mathcal{N}^\vee$. So we can assume that the last rule in the derivation is a proper rule and consider all possible cases of such a rule. The cut rule can be used, because in every case we are dealing with compatible sequents.

Of the left rules only three, namely, **NG-L**, **CD-L** and **ULQ-L**, present some problems, because there is a right component that in some cases is not local, although by regularity we can assume it is a Π-formula.

Rule **NG-L**. The last rule in the derivation is of the form,

$$\mathcal{M} \vdash \mathcal{N}, \phi \Rightarrow \neg\phi, \mathcal{M} \vdash \mathcal{N}.$$

We know that $\neg\phi$ is regular, so ϕ is Π-formula. First, we consider the case where ϕ is local. If $\mathcal{N}$ is empty we apply the induction hypothesis. If $\mathcal{N}$ is not empty we have the following derivations in the intuitionistic calculus:

(1) $\mathcal{M}', \mathcal{M} \vdash \mathcal{N}^{\vee} \vee \phi$ (induction hypothesis).

(2) $\mathcal{M}', \neg\phi, \mathcal{M} \vdash \mathcal{N}^{\vee}$ (Corollary 10.3.3).

If ϕ is not local, then ϕ is derived by global universal quantification from a local formula ϕ'. To simplify the notation we write just one quantifier, but the argument is similar if there are more quantifiers. So, we set $\phi = (\forall Y)\phi'(Y)$. From Theorem 9.2 (or Corollary 9.2.1) there is a derivation in the classical calculus for $\mathcal{M} \vdash \mathcal{N}, \phi(Y/Z)$ of height $k' \leq k$, where Z is a new variable and this sequent is regular, so we can apply the induction hypothesis. If $\mathcal{N}$ is empty, then by the induction hypothesis we have $\mathcal{M} \vdash \phi'(Y/Z)$ derivable in the intuitionistic calculus, so we apply rules **UGQ-R** and **NG-L**. If $\mathcal{N}$ is non-empty we have the following derivations:

(1) $\mathcal{M}', \mathcal{M} \vdash \mathcal{N}^{\vee} \vee \phi'(Y/Z)$ (induction hypothesis).

(2) $\mathcal{M}', \mathcal{M}, \neg\mathcal{N}^{\vee} \vdash \phi'(Y/Z)$ (Corollary 10.3.3).

(3) $\mathcal{M}', \mathcal{M}, \neg\mathcal{N}^{\vee} \vdash (\forall Y)\phi'(Y)$ (rule **UGQ-R**).

(4) $\neg(\forall Y)\phi'(Y), \mathcal{M}', \mathcal{M} \vdash \neg\neg\mathcal{N}^{\vee}$ (rules **NG-L** and **NG-R**).

(5) $\mathcal{M}', \mathcal{N}^{\vee} \vee \neg\mathcal{N}^{\vee}, \neg(\forall Y)\phi'(Y), \mathcal{M} \vdash \mathcal{N}^{\vee}$ (Corollary 10.3.2).

Rule **CD-L**. The last rule in the derivation is of the form,

$$\mathcal{M} \vdash \mathcal{N}, \phi_1; \phi_2, \mathcal{M} \vdash \mathcal{N} \Rightarrow \phi_1 \rightarrow \phi_2, \mathcal{M} \vdash \mathcal{N}.$$

By regularity we know that ϕ_1 is Π-formula and ϕ_2 is regular. We have two cases, as in rule **NG-L**. In the first case we assume the formula ϕ_1 is local.

The case where $\mathcal{N}$ is empty follows from the induction hypothesis. If $\mathcal{N}$ is not empty we have the following derivations:

(1) $\mathcal{M}', \mathcal{M} \vdash \mathcal{N}^{\vee} \vee \phi_1$ (induction hypothesis).

(2) $\mathcal{M}', \phi_2, \mathcal{M} \vdash \mathcal{N}^{\vee}$ (induction hypothesis).

(3) $\mathcal{M}', \mathcal{M}, \neg\mathcal{N}^{\vee} \vdash \phi_1$ (Corollary 10.3.3)

(4) $\neg\mathcal{N}^{\vee}, \phi_1 \rightarrow \phi_2, \mathcal{M}', \mathcal{M} \vdash \mathcal{N}^{\vee}$ (rule **CD-L**).

(5) $\phi_1 \rightarrow \phi_2, \mathcal{M}', \mathcal{M} \vdash \neg\neg\mathcal{N}^{\vee}$(rules **NG-L**, **INC**, and **NG-R**).

(6) $\mathcal{N}^{\vee} \vee \neg\mathcal{N}^{\vee}, \neg\mathcal{M}', \phi_1 \rightarrow \phi_2, \mathcal{M} \vdash \mathcal{N}^{\vee}$ (Corollary 10.3.2).

If the first case fails we assume ϕ_1 is of the form $(\forall Y)\phi'(Y)$, where $\phi'(Y)$ is local, and by Theorem 9.2 (or Corollary 9.2.1), there is a derivation for $\mathcal{M} \vdash \mathcal{N}, \phi'(Y/Z)$ of height $k' \leq k$, where Z is a new variable. The case where $\mathcal{N}$ is empty follows by the induction hypothesis and rules **UGQ-R**, and **CD-L**. If $\mathcal{N}$ is not empty we have the following derivations:

(1) $\mathcal{M}', \mathcal{M} \vdash \mathcal{N}^{\vee} \vee \phi'(Y/Z)$ (induction hypothesis).

(2) $\mathcal{M}', \phi_2, \mathcal{M} \vdash \mathcal{N}^{\vee}$ (induction hypothesis).

(3) $\neg\mathcal{N}^{\vee}, \mathcal{M}', \mathcal{M} \vdash \phi'(Y/Z)$ (Corollary 10.3.3)

(4) $\neg\mathcal{N}^{\vee}, \mathcal{M}', \mathcal{M} \vdash (\forall Y)\phi'(Y)$ (rule **UGQ-R**).

(5) $\neg\mathcal{N}^{\vee}, \mathcal{M}', \phi_1 \rightarrow \phi_2, \mathcal{M} \vdash \mathcal{N}^{\vee}$ (rule **CD-L**).

(6) $\mathcal{M}', \phi_1 \rightarrow \phi_2, \mathcal{M} \vdash \neg\neg\mathcal{N}^{\vee}$ (rules **NG-L**, **INC**, and **NG-R**).

(7) $\neg\mathcal{N}^{\vee} \vee \mathcal{N}^{\vee}, \mathcal{M}', \phi_1 \rightarrow \phi_2, \mathcal{M} \vdash \mathcal{N}^{\vee}$ (Corollary 10.3.2).

Rule **ULQ-L** is similar to **CD-L** when ϕ_1 is local, and it is handled in the same way using Corollaries 10.3.2 and 10.3.3. All the other left rules follow immediately from the induction hypothesis.

In general, the right rules require the induction hypothesis and some manipulation. We discuss some examples, noting that by regularity the global quantifier rules are impossible.

Rule **NG-R**. The last rule in the derivation is of the form,

$$\phi, \mathcal{M} \vdash \mathcal{N} \Rightarrow \mathcal{M} \vdash \mathcal{N}, \neg\phi.$$

Note that the formula ϕ is local. The case where $\mathcal{N}$ is empty is trivial. Otherwise, we have the following derivations in the intuitionistic calculus:

(1) $\phi, \mathcal{M}', \mathcal{M} \vdash \mathcal{N}^{\vee} \vee \neg\phi$ (induction hypothesis and rule **DJ-R**).

(2) $\neg\phi \vdash \mathcal{N}^{\vee} \vee \neg\phi$ (rule **DJ-R**).

(3) $\phi \vee \neg\phi, \mathcal{M}', \mathcal{M} \vdash \mathcal{N}^{\vee} \vee \neg\phi$ (rule **DJ-L**).

Rule **CD-R**. The last rule in the derivation is of the form,

$$\phi_1, \mathcal{M} \vdash \mathcal{N}, \phi_2 \Rightarrow \mathcal{M} \vdash \mathcal{N}, \phi_1 \to \phi_2.$$

Note that ϕ_1 and ϕ_2 are local. The case $\mathcal{N}$ is empty is trivial. Otherwise, we have the following derivations:

(1) $\mathcal{M}', \phi_1, \mathcal{M} \vdash \mathcal{N}^{\vee} \vee \phi_2$ (induction hypothesis).

(2) $\mathcal{M}', \phi_1, \mathcal{M}, \neg\mathcal{N}^{\vee} \vdash \phi_2$ (Corollary 10.3.3).

(4) $\mathcal{M}', \mathcal{M}, \neg\mathcal{N}^{\vee} \vdash \mathcal{N}^{\vee} \vee \phi_1 \to \phi_2$ (rules **CD-R** and **DJ-R**).

(5) $\mathcal{N}^{\vee} \vdash \mathcal{N}^{\vee} \vee \phi_1 \to \phi_2$.

(6) $\mathcal{N}^{\vee} \vee \neg\mathcal{N}^{\vee}, \mathcal{M}', \mathcal{M} \vdash \mathcal{N}^{\vee} \vee \phi_1 \to \phi_2$ (rule **DJ-L**).

Rule **ULQ-R**. The last rule in the derivation is of the form,

$$X \in W, \mathcal{M} \vdash \mathcal{N}, \phi(Y/X) \Rightarrow \mathcal{M} \vdash \mathcal{N}, (\forall Y \in W)\phi(Y).$$

The case where $\mathcal{N}$ is empty follows from the induction hypothesis. Otherwise, we have the following derivations:

(1) $\mathcal{M}', X \in W, \mathcal{M} \vdash \mathcal{N}^{\vee} \vee \phi(Y/X)$ (induction hypothesis).

(2) $X \in W, \mathcal{M}', \mathcal{M}, \neg\mathcal{N}^{\vee} \vdash \phi(Y/X)$ (Corollary 10.3.3).

(3) $\neg\mathcal{N}^{\vee}, \mathcal{M}', \mathcal{M} \vdash \mathcal{N}^{\vee} \vee (\forall Y \in W)\phi(Y)$ (rule **ULQ-R**).

(4) $\mathcal{N}^{\vee} \vee \neg\mathcal{N}^{\vee}, \mathcal{M}', \mathcal{M} \vdash \mathcal{N}^{\vee} \vee (\forall Y \in W)\phi(Y)$ (rule **DJ-L**).

Rule **DJ-R**. The last rule in the derivation is of the form,

$$\mathcal{M} \vdash \mathcal{N}, \phi \Rightarrow \mathcal{M} \vdash \mathcal{N}, \phi \vee \psi.$$

The case where $\mathcal{N}$ is empty follows from the induction hypothesis. Otherwise, we have the following derivations:

(1) $\mathcal{M}', \mathcal{M} \vdash \mathcal{N}^{\vee} \vee \phi$ (induction hypothesis).

(2) $\mathcal{N}^{\vee} \vdash \mathcal{N}^{\vee} \vee (\phi \vee \psi)$ (rule **DJ-R**).

(3) $\phi \vdash \mathcal{N}^{\vee} \vee (\phi \vee \psi)$ (rule **DJ-R**).

(4) $\mathcal{N}^{\vee} \vee \phi \vdash \mathcal{N}^{\vee} \vee (\phi \vee \psi)$ (rule **DJ-R**).

(5) $\mathcal{M}', \mathcal{M} \vdash \mathcal{N}^{\vee} \vee (\phi \vee \psi)$ (cut rule).

The other rules are left to the reader. □

Corollary 10.4.1 If ϕ is a Π-formula that is a formal theorem in classical logic, then ϕ is a formal theorem in intuitionistic logic.

PROOF. If ϕ is local, this follows from Theorem 10.4. Otherwise, ϕ is obtained by universal quantification from a local formula. For example, let ϕ be the formula $(\forall Y)\phi'$. From Corollary 9.2.1 it follows that $\phi'(Y/Z)$ is also a formal theorem in classical logic, hence by Theorem 10.4 it follows that $\phi'(Y/Z)$ is a formal theorem in intuitionistic logic. Using rule **UGQ-R** we conclude that ϕ is a formal theorem in intuitionistic logic. □

Corollary 10.4.2 *Let ϕ be a formal theorem in classical logic of the form $\phi_1 \to \phi_2$, where ϕ_1 is regular and ϕ_2 is a Π-formula. Then ϕ is a formal theorem in intuitionistic logic.*

PROOF. The assumption means that $\mathcal{M} \vdash \phi$ is derivable in the classical calculus, where the components of $\mathcal{M}$ are closed axioms. It follows that $\mathcal{M}, \phi_1 \vdash \phi_2$ is derivable in the classical calculus, and by Theorem 10.4 it is derivable in the intuitionistic calculus. We conclude that ϕ is a formal theorem in intuitionistic logic. □

EXAMPLE 10.3 Corollary 10.4.2 applies in particular to Theorem 3.3, where ϕ is of the form $(\forall Z)\phi' \to (\forall Z)\phi''$, and ϕ', ϕ'' are local formulas (see Example 9.2). It follows that Theorem 3.3 is a formal theorem in intuitionistic logic. A proof by induction consists in proving $(\forall Z)\phi'$, which is Π-formula. By Corollary 10.4.1, such a proof is valid in intuitionistic logic. We conclude that all theorems we have proved by induction are actually formal theorems in intuitionistic logic. □

10.3 Local Reduction

In this section we extend the reduction technique of Chapter 9 to derivations in the intuitionistic calculus.

Theorem 10.5 *Let χ be a local reduction of ϕ. If Z is a variable that does not occur free in χ, then there is a local reduction χ' of the form $(\exists Z \in W)\phi'(Y/Z)$, and $\chi \to \chi'$ is a formal theorem in intuitionistic logic.*

PROOF. Same proof as for Theorem 9.5, noting that the derivation there can be formalized in the intuitionistic calculus. □

If ϕ is a formula of the form $(\exists Y)\phi'$ we say that ϕ is an *existential formula*. Note that we do not impose any restriction on the formula ϕ'. If ϕ

is a formula where no sub-formula is existential, we say that ϕ is *existential-free*. Note that a Π-formula is existential-free and, in particular, the closed axioms are existential-free.

Theorem 10.6 *Assume there is a derivation in the intuitionistic calculus for a sequent $\mathcal{M} \vdash \phi$ where all the components of $\mathcal{M}$ are existential-free formulas and ϕ is existential. Then there is a derivation in the intuitionistic calculus for a sequent $\mathcal{M} \vdash \chi$ where χ is a local reduction of ϕ.*

PROOF. Note that the only way to introduce ϕ in the right is by rule **EGQ-R** in the form,

$$\mathcal{M}' \vdash \phi'(Y/V) \Rightarrow \mathcal{M}' \vdash (\exists Y)\phi'(Y).$$

The derivation may contain several occurrences of similar applications of rule **EGQ-R**, and the term V may change from one application to another. Since the components of $\mathcal{M}$ are existential-free, it is clear that once ϕ has been introduced as a right component in a branch of the derivation, it remains there until the end of the derivation, hence from that point the branch contains no application of a rule involving ϕ. In fact, only left rules occur in the branch, and the only left rule which is excluded is **ELQ-L**, because it introduces a global existential quantifier. Still, note that this rule is excluded in the branch only after ϕ has been introduced as a right component.

We intend to cancel all the applications of rule **EGQ-R** where the formula ϕ is introduced, so in the example above the formula ϕ is replaced by $\phi'(Y/V)$, which is a local reduction of ϕ. We now try to complete the derivation in such a way that at the bottom of the derivation we have a sequent $\mathcal{M} \vdash \chi$, where χ is a local reduction of ϕ. The general procedure is to replace every right occurrence of ϕ by the preceding local reduction. In most cases rules are preserved after this change, because the formula ϕ is replaced by the same formula before and after the application of the rule. In fact, only two rules are affected by the change, and we proceed to discuss the procedure in each case.

In the first case we have an application of rule **DJ-L**, that introduces a disjunction on the left. Originally, the application was of the form,

$$\psi_1, \mathcal{M}_1 \vdash \phi; \psi_2, \mathcal{M}_1 \vdash \phi \Rightarrow \Psi_1 \vee \psi_2, \mathcal{M}_1 \vdash \phi,$$

but now we find the ϕ has been replaced by a local reduction χ_1 in one derivation, and by χ_2 in the other derivation. So we have the following derivations:

(1) $\psi_1, \mathcal{M}_1 \vdash \chi_1$.

(2) $\psi_2, \mathcal{M}_1 \vdash \chi_2$.

(3) $\psi_1, \mathcal{M}_1 \vdash \chi_1 \vee \chi_2$ (rule **DJ-R** (*a*)).

(4) $\psi_2, \mathcal{M}_1 \vdash \chi_1 \vee \chi_2$ (rule **DJ-R** (*b*)).

(5) $\psi_1 \vee \psi_2, \mathcal{M}_1 \vdash \chi_1 \vee \chi_2$ (rule **DJ-L**).

So, at this stage the formula ϕ must be replaced by the local reduction $\chi_1 \vee \chi_2$.

The second case is more delicate, as it involves an application of rule **ELQ-L**, which imposes some restrictions via the characteristic variable, and we may find that the characteristic variable X occurs free in the local reduction that has replaced ϕ. The original application of the rule is

$$X \in W', \psi(Y'/X), \mathcal{M}_1 \vdash \phi \Rightarrow (\exists Y' \in W')\psi, \mathcal{M}_1 \vdash \phi,$$

and after the replacement it has the form,

$$X \in W', \psi(Y'/X), \mathcal{M}_1 \vdash \chi \Rightarrow (\exists Y' \in W')\psi, \mathcal{M}_1 \vdash \chi,$$

where χ is a local reduction of ϕ. If the characteristic variable X does not occur free in χ, this is still a correct application of rule **ELQ-L**, so there is no problem. If this is not the case, we fix the derivation as follows:

(1) $X \in W', \psi(Y'/X), \mathcal{M}_1 \vdash \chi(X/X)$.

(2) $X \in W', \psi(Y'/X), \mathcal{M}_1 \vdash X \in W'$ (initial sequent and inclusion).

(3) $X \in W', \psi(Y'/X), \mathcal{M}_1 \vdash (\exists X \in W')\chi(X)$ (rule **ELQ-R**).

(4) $(\exists Y' \in W')\psi, \mathcal{M}_1 \vdash (\exists X \in W')\chi(X)$ (rule **EGQ-L**).

The new local reduction that replaces ϕ at this stage of the derivation is $(\exists X \in W')\chi(X)$.

Following this procedure we get a derivation for a sequent $\mathcal{M} \vdash \chi$, where χ is a local reduction of ϕ. □

Corollary 10.6.1 *If ϕ is a formal theorem in intuitionistic logic there is a local reduction χ' of ϕ of the form $(\exists Z \in W)\phi'(Y/Z)$ that is also a formal theorem in intuitionistic logic.*

PROOF. Immediate from Theorems 10.5 and 10.6, noting the closed axioms are existential-free assertions. □

EXERCISE 10.14 Assume a formula of the form $\psi \rightarrow \phi$ is a formal theorem in intuitionistic logic, where ψ is existential-free, and ϕ is of the form $(\exists Y)\phi'$. Prove that there is local reduction χ' of ϕ of the form $(\exists Z \in W)\phi'(Y/Z)$ such that $\psi \rightarrow \chi'$ is also a formal theorem in intuitionistic logic.

Remark 10.3 The discussion in Remark 9.5 also applies here, because the local reduction theorems are similar. In particular, the selector operation f can be introduced under the same assumptions. Still, there are some important differences. Theorem 10.6 appears to be stronger than Theorem 9.4 because in the latter the formula ϕ is required to be a Σ-formula, while in the former we require only that ϕ is existential. Still, in applications of Theorem 10.6 we need ϕ to be of the form $(\exists Y)\phi'$ with ϕ' local, otherwise we cannot introduce the predicate p of Remark 9.5. But Theorem 10.6 is stronger in the sense that the components of $\mathcal{M}$ are required to be existential-free, while in Theorem 9.4 we require that they must be Π-formulas. In particular, in the application of Remark 9.5 we may assume that the formula ψ is existential-free. □

10.4 Notes

Although we take the official position that intuitionistic logic provides the right support for the theory of sets, the real situation is much more liberal, since we allow classical logic in dealing with sets in the universe. We require intuitionistic logic only in dealing with the universe via global quantifiers. The classical reduction theorem shows that the use of classical logic to prove local formulas is consistent with the use of intuitionistic logic.

This situation depends essentially on the form of the axioms and might change if the theory is extended with new axioms that are not regular formulas. For example, an axiom of the form $(\forall X)(\exists Y)\phi$ would be beyond Theorem 10.4, although we may expect that the existential quantifier could be avoided by means of a set operation.

The consideration of intuitionistic logic in this work is not related to attempts at a constructive set theory, where classical logic is altogether rejected. We are forced to bring in intuitionistic logic because of the peculiar nature of the universe of sets, and we rely on classical logic as long as we deal with the local universe. This is not a concession to any particular philosophy or ideology, but rather a reaction to an objective situation.

A complete formulation of a system of constructive set theory, with references, is given in Myhill (1975).

Appendix A

Enumeration

The ordinal permutation rule in Chapter 8 is a crucial element in the system $\mathbb{G}$, for it induces the power-set construction, the theory of cardinals, and the axiom of choice. On the other hand, the rule presents several conceptual difficulties, for it involves the class of all ordinals, which is not a set, although this class occurs only in a potential way, and for a given set Z only a segment of the ordinals is required. Still, the construction is not as transparent as, for example, the fundamental sequence that supports the omega rule.

The ordinal permutation rule is essential for the proper formulation of the theory of sets if we want to preserve the main themes derived from Cantor's formulation. Still, it appears that it could be of some interest to introduce a new rule that is weaker but more transparent. This is the ordinal enumeration rule, defined below, which can be used to derive a substantial segment of classical set theory. This rule is actually a consequence of the ordinal permutation rule. In this Appendix we assume only material preceding Chapter 8, in particular the theory of ordinals in Chapter 6 and the omega rule in Chapter 7.

We present several applications of the ordinal enumeration rule, including a new explicit version of the set induction rule, where the argument of Section 3.2 via induction trees is formalized in terms of ordinal enumerations. Still, this application assumes the inductive and recursive properties of the natural numbers. Furthermore, the construction does not allow for the elimination of the set recursion rule.

All the results in this Appendix belong to the system $\mathbb{G}$, since the enumeration rule can be derived from the ordinal permutation rule (see Remark A.5).

A.1 Ordinal Enumerations

An important application of the set ω is the possibility of enumerating elements of a set Z, not necessarily all of them. In principle such an enumeration takes the form of a sequence $f(0), f(1), ..., f(\mathsf{n}), ...$ where all the components belong to the set Z, and f is a given unary set operation. Using ω this sequence becomes a set $S = (\lambda \mathsf{n} \in \omega) f(\mathsf{n})$. Once this construction is available we can ask questions about sets that are *denumerable*, and even about all possible enumerations of a given set Z.

We can generalize this idea and introduce α-enumerations in a set Z, where α is a given ordinal. Such an enumeration takes the form $f(0)$, $f(1)$,...,$f(\beta)$, ... where $\beta <_{\mathbf{or}} \alpha$. Again, via local abstraction the enumeration becomes a function $F = (\lambda\beta <_{\mathbf{or}} \alpha) f(\beta)$. In this context we can talk about sets that are α-denumerable, etc.

A.1.1 $\mathbf{EN}(Z, F) \equiv \mathbf{FU}(F) \wedge \mathbf{OR}(\mathbf{do}(F)) \wedge \mathbf{ra}(F) \subseteq Z$ (*F is an ordinal enumeration in Z*).

If F is an ordinal enumeration in Z and $\mathbf{do}(F) = \alpha$, we shall say that F is an *α-enumeration in Z*. If F is an ordinal enumeration in Z where $\mathbf{do}(F) = \alpha$ and $\mathbf{ra}(F) = Z$, we say that F is an *ordinal enumeration (α-enumeration) of Z*.

EXERCISE A.1 Prove that if $X \subseteq Z$ and $\mathbf{EN}(X, F)$, then $\mathbf{EN}(Z, F)$.

EXAMPLE A.1 Let F be a function where $\mathbf{do}(F) = \omega +_{\mathsf{or}} \omega$, defined as follows: if $\mathsf{n} <_{\mathsf{or}} \omega$, then $\mathbf{ap}(F, \mathsf{n}) = 2 \times \mathsf{n}$ and $\mathbf{ap}(F, \omega +_{\mathsf{or}} \mathsf{n}) = (2 \times \mathsf{n}) + 1$. Then, F is a 1–1 ordinal enumeration of ω. □

We intend to introduce a set operation $\Upsilon(Z, \alpha)$ that contains all the α-enumerations in the set Z. We approach this construction via the notion of *free choice α-sequence* in Z. This is a process where for each ordinal $\beta <_{\mathbf{or}} \alpha$ a set $X_\beta \in Z$ is chosen, the choice being arbitrary or executed under some restrictions. As in the case of ordinal permutations we assume that all choices are activated (almost) simultaneously, but the actual choice of X_β takes place only after all choices of X_γ for $\gamma <_{\mathbf{or}} \beta$ have been executed. On the other hand, we do not require that the choice X_β be different from the preceding choices.

This type of process is similar, to some extent, to the free choice ordinal permutations of Chapter 8. The main difference is that the ordinal α is fixed, so the class of all ordinals is not invoked. For this reason there is no need of a principle of closure. Furthermore, we allow repetitions, which were excluded in Chapter 8, precisely to ensure the validity of the principle of closure. On the other hand, the condition that the choice X_β takes place after all preceding choices have been executed is essential for the validity of the completeness axiom below.

Here again we move to a more general process where all the free choice α-sequences in Z are generated simultaneously. The completion of this general process is the set of all α-enumerations in Z. This is an objective primitive construction that supports the enumeration rule below.

Remark A.1 In the preceding discussion we took the position that the free choice α-sequences in Z are a simplified form of the more general processes considered in Chapter 8. If $\alpha = \omega$ the situation is still simpler, because we can assume that the choices take place one after the other, rather than simultaneously as proposed above. In this way a free choice ω-sequence appears to be a generalization of the fundamental sequence that supports the introduction of ω in Chapter 7. Furthermore, the general process that generates all the ω-sequences in Z can be described as a tree where at each node every element of Z can be taken as a successor. Thus we can consider $\Upsilon(Z,\omega)$ as the completion of the tree. In this way we get a more transparent construction for the set operation $\Upsilon(Z,\omega)$. □

Rule of Ordinal Enumeration We introduce a binary set operation Υ, the *enumeration operation*, that satisfies the following two axioms.

The Consistency Axiom:

$$F \in \Upsilon(Z,\alpha) \to \mathbf{EN}(Z,F) \wedge \mathbf{do}(F) = \alpha.$$

The Completeness Axiom: If p is a general binary predicate, and $Z \neq \emptyset$, there exists $F \in \Upsilon(Z,\alpha)$ such that

$$(\forall \beta <_{\mathbf{or}} \alpha)((\exists Y \in Z)p(F\restriction\beta, Y) \to p(F\restriction\beta, \mathbf{ap}(F,\beta))).$$

Remark A.2 The condition $Z \neq \emptyset$ in the completeness axiom is necessary, for if $0 <_{\mathbf{or}} \alpha$, then $\Upsilon(\emptyset,\alpha) = \emptyset$. On the other hand, $\Upsilon(\emptyset,0) = \{\emptyset\}$, and the completeness axiom holds. In this section the first argument of the operation Υ is always assumed to be non-empty. □

Remark A.3 The operation Υ has been introduced under the assumption that the second argument is an ordinal α. On the other hand, we know that set operations are intended to be defined for arbitrary sets in the universe. To complete the definition we stipulate that in case X is not an ordinal, then $\Upsilon(Z,X) = \emptyset$. The same convention is implicitly assumed for several set operations, to be defined later, where some of the arguments are intended to be ordinals: whenever one of the intended arguments is not an ordinal, the value of the operation is $\emptyset$. We have a similar situation concerning predicates, where some of the arguments are intended to be ordinals. The convention here is that whenever one of the intended arguments is not an ordinal, then the predicate is false. □

Remark A.4 The completeness axiom contains a general form (in a local context) of the axiom of dependent choices. □

Remark A.5 The ordinal enumeration rule can be derived from the ordinal permutation rule, so it is part of the system $\mathbb{G}$. By means of replacement with the power-set operation we define,

$$\Upsilon(Z, \alpha) = \{F \in \mathbf{pw}(\alpha \times Z) :: \mathbf{EN}(Z, F) \wedge \mathbf{do}(F) = \alpha\}.$$

It is clear that the consistency axiom follows immediately from this definition. To prove the completeness axiom, let p be a choice predicate. We want an ordinal enumeration F in Z such that whenever $\beta <_{\mathbf{or}} \alpha$ and $(\exists Y \in Z)p(F{\restriction}\beta, Y)$, then $p(F{\restriction}\beta, \mathbf{ap}(F, \beta))$ holds. Let p' be the following binary predicate:

$$p'(V, Y) \equiv [Y]_1 = \mathbf{do}(V) \wedge (p(\mathbf{ra}(V), [Y]_2) \vee (\forall Y' \in Z)\neg p(\mathbf{ra}(V), Y')).$$

Let $F' \in \Xi(\alpha \times Z)$ be an ordinal permutation induced by the predicate p'. We set $F = F'[\alpha]$ and it follows that F is the α-enumeration in Z induced by the choice predicate p. □

EXERCISE A.2 Let F and F' be the functions from Remark A.5. Prove that the following properties hold for any $\beta <_{\mathbf{or}} \alpha$:

(a) $\alpha \leq_{\mathbf{or}} \mathbf{do}(F')$ and $\mathbf{ap}(F', \beta) = <\beta, Y>$, where $Y \in Z$, and if there exists $Y' \in Z$ such that $p(\mathbf{ra}(F'{\restriction}\beta), Y')$, then $p(\mathbf{ra}(F'{\restriction}\beta), Y)$.

(b) F is a function, $\mathbf{do}(F) = \alpha$, and $\mathbf{ra}(F'{\restriction}\beta) = F{\restriction}\beta$.

(c) $\mathbf{ap}(F, \beta) = Y$ where $Y \in Z$, and if there exists $X \in Z$ such that $p(F{\restriction}\beta, X)$, then $p(F{\restriction}\beta, Y)$.

Theorem A.1 *Let F be an α-enumeration in Z. Then $F \in \Upsilon(Z, \alpha)$.*

PROOF. Consider the choice predicate $p(V, Y) \equiv \mathbf{ap}(F, \mathbf{do}(V)) = Y$, and let F' be an element of $\Upsilon(Z, \alpha)$ induced by the completeness axiom. Clearly, $\mathbf{ap}(F, \beta) = \mathbf{ap}(F', \beta)$ whenever $\beta <_{\mathbf{or}} \alpha$, hence $F = F'$, and $F \in \Upsilon(Z, \alpha)$. □

EXERCISE A.3 Prove that $\Upsilon(Z,\alpha) = \emptyset$ if and only if $Z = \emptyset$ and $0 <_{\mathbf{or}} \alpha$.

EXERCISE A.4 Prove that if $X \subseteq Z$, then $\Upsilon(X,\alpha) \subseteq \Upsilon(Z,\alpha)$.

A.2 Countable Sets

We can use ordinal enumeration to define several classical set theoretical notions. We start with the formal definitions, and later prove a number of typical properties.

A.2.1 $\mathbf{CT}(Z,\alpha) \equiv Z = \emptyset \vee (\exists F \in \Upsilon(Z,\alpha))\mathbf{ra}(F) = Z$ (*Z is α-countable*).

A.2.2 $\mathbf{DE}(Z,\alpha) \equiv (\exists F \in \Upsilon(Z,\alpha))\mathbf{FU1}(F) \wedge \mathbf{ra}(F) = Z$ (*Z is α-denumerable*).

A.2.3 $\mathbf{cpw}(Z,\alpha) = \{\mathbf{ra}(F) : F \in \Upsilon(Z,\alpha) :\} \cup \{\emptyset\}$ (*the set of α-countable subsets of Z*).

Clearly, if Z is α-denumerable, then Z is α-countable.

Remark A.6 The composition and inversion of functions play an important role in the theory of α-enumerations. These are general constructions that were informally introduced in Chapter 8. If G is a β-enumeration of α and F is an α-enumeration of Z, then the composition of G and F, denoted by $G \circ F$, is a β-enumeration of Z. Equivalently, if Z is α-countable and α is β-countable, then Z is β-countable. Note that if G and F are 1–1 functions, then $G \circ F$ is also a 1–1 function.

We recall also that the inverse of a 1–1 function is a 1–1 function. Hence, if F is a 1–1 α-enumeration of β, then the inverse of F is a β-enumeration of α. It follows that when β is α-denumerable, then α is β-denumerable. □

EXAMPLE A.2 Let F be a function where $\mathbf{do}(F) = \alpha$ and $\mathbf{ra}(F) = Z$. It follows that $F \in \Upsilon(Z,\alpha)$ and Z is α-countable. If F is 1–1, then Z is α-denumerable. Note that $\mathbf{cpw}(Z,0) = \mathbf{cpw}(\emptyset,\alpha) = \{\emptyset\}$ holds for any set Z

and any ordinal α. In particular, if $Z = \alpha$ we can take $F = \{<\gamma, \gamma> : \gamma <_{\mathbf{or}} \alpha :\}$ Clearly, F is a 1–1 function, where $\mathbf{do}(F) = \alpha$ and $\mathbf{ra}(F) = \alpha$, so F is α-enumeration of α. We conclude that α is α-denumerable. □

EXERCISE A.5 Let F be a function such that $\mathbf{do}(F)$ is α-countable. Prove that $\mathbf{ra}(F)$ is α-countable.

EXERCISE A.6 Prove that if $\alpha \leq_{\mathbf{or}} \beta$, then α is β-countable.

EXERCISE A.7 Prove that if $X \subseteq Z$ and Z is α-countable, then X is α-countable.

EXERCISE A.8 Let F be a 1–1 function where $\mathbf{do}(F) = Z$ and $\mathbf{ra}(F) \subseteq \alpha$. Prove that Z is α-countable.

Theorem A.2 *Let Z and X be arbitrary sets. The following conditions are equivalent:*

(i) $X \in \mathbf{cpw}(Z, \alpha)$.

(ii) $X \subseteq Z \wedge \mathbf{CT}(X, \alpha)$.

PROOF. The case $X = \emptyset$ is trivial, so we assume $X \neq \emptyset$. If (i) holds there exists $F \in \Upsilon(Z, \alpha)$ such that $\mathbf{ra}(F) = X$. It follows that $X \subseteq Z$ and $F \in \Upsilon(X, \alpha)$, so X is α-countable. Conversely, if (ii) holds there is $F \in \Upsilon(X, \alpha)$ where $\mathbf{ra}(F) = X$. It follows that $F \in \Upsilon(Z, \alpha)$, so $X \in \mathbf{cpw}(Z, \alpha)$. □

Corollary A.2.1 *Assume that Z is α-countable and that X is an arbitrary set. The following conditions are equivalent:*

(i) $X \in \mathbf{cpw}(Z, \alpha)$.

(ii) $X \subseteq Z$.

PROOF. The property is trivial if $X = \emptyset$, so we assume that $X \neq \emptyset$. The implication from (i) to (ii) follows from Theorem A.2. If $X \subseteq Z$ and Z is α-countable, there exists $F \in \Upsilon(Z, \alpha)$, where $\mathbf{ra}(F) = Z$, so we take $Y \in X$ and redefine F to F', where $\mathbf{do}(F') = \alpha$, $\mathbf{ap}(F', \beta) = \mathbf{ap}(F, \alpha)$ when $\mathbf{ap}(F, \beta) \in X$, and $\mathbf{ap}(F', \beta) = Y$ when $\mathbf{ap}(F, \beta) \notin X$. It follows that $F' \in \Upsilon(Z, \alpha)$ and $\mathbf{ra}(F') = X$, so $X \in \mathbf{cpw}(Z)$. □

EXERCISE A.9 Assume that Z is a set and that α is an ordinal. Prove that the following conditions are equivalent:

(a) Z is α-countable.

(b) $Z \in \mathbf{cpw}(Z, \alpha)$.

Theorem A.3 *Assume that Z is α-countable and X is a set. Then:*

(i) *If $X \subseteq Z$, then X is α-countable.*

(ii) *$Z \setminus X$ is α-countable.*

(iii) *If $Z \in X$, then $\bigcap X$ is α-countable.*

PROOF. Part (i) follows from Theorem A.2 and Corollary A.2.1. Parts (ii) and (iii) follow from (i), noting that $Z \setminus X \subseteq Z$, and $\bigcap X \subseteq Z$. □

Remark A.7 Theorem A.2 implies that $\mathbf{cpw}(Z, \alpha)$ is the set of all subsets of Z that are α-countable. On the other hand, if Z is α-countable, then $\mathbf{cpw}(Z, \alpha)$ is the power-set of Z. In particular, $\mathbf{cpw}(\omega, \omega)$ is the power-set of ω, and in general $\mathbf{cpw}(\alpha, \alpha)$ is the power-set of α. We do not know of any ordinal α such that, under the ordinal enumeration rule, $\mathbf{cpw}(\omega, \omega)$ is α-countable, although such an ordinal exists under the ordinal permutation rule of Chapter 8. □

EXERCISE A.10 Assume that the set Z is α-denumerable. Prove that $\mathbf{cpw}(Z, \alpha)$ is not α-countable. □

Theorem A.4 *Let Z be a set and α an ordinal. The following conditions are equivalent:*

(i) $\mathbf{NT}(\alpha) \wedge \mathbf{DE}(Z, \alpha)$.

(ii) $\mathbf{FI}(Z) \wedge \mathbf{ct}(Z) = \alpha$.

PROOF. The implication from (i) to (ii) follows from Theorem 5.6. In the other direction we use Theorem 5.4 (i) and Example 5.4. □

Theorem A.5 *Assume that Z is α-countable (α-denumerable), where α is an infinite ordinal. Then $S(Z)$ is α-countable (α-denumerable).*

PROOF. The result is trivial if $Z = S(Z)$, so we assume $Z \notin Z$. Let F be a function where $\mathbf{do}(F) = \alpha$ and $\mathbf{ra}(F) = Z$. We define a function F' where $\mathbf{do}(F') = \alpha$ and $\mathbf{ra}(F') = S(Z)$. We set $\mathbf{ap}(F', 0) = Z$, $\mathbf{ap}(F', S(\mathsf{n})) = \mathbf{ap}(F, \mathsf{n})$ whenever $0 \leq \mathsf{n} < \omega$, and $\mathbf{ap}(F', \beta) = \mathbf{ap}(F, \beta)$ whenever $\omega \leq_{\mathbf{or}} \beta$. It follows that $S(Z)$ is α-countable. If Z is α-denumerable, then the function F is 1–1, and it follows that F' is also 1–1. □

EXERCISE A.11 Prove that any finite set is ω-countable.

Several results in this section are proved under the assumption that α is an ordinal such that $\alpha \times \alpha$ is α-countable. The only finite ordinals that satisfy this condition are 0 and $1 = \{0\}$. We know that every critical ordinal κ satisfies a stronger condition: the set $\kappa \times \kappa$ is κ-denumerable (Theorem 7.6). In fact, ω is a critical ordinal and $\omega \times \omega$ is ω-denumerable. We shall show later that every infinite ordinal satisfies the same condition.

Theorem A.6 *Assume that $\alpha \times \alpha$ is α-countable (α-denumerable) and that the sets X and Y are α-countable (α-denumerable). Then $X \times Y$ is α-countable (α-denumerable).*

PROOF. If G is an α-enumeration of $\alpha \times \alpha$, F_1 is an α-enumeration of X, and F_2 is an α-enumeration of Y, then we get an α-enumeration F of $X \times Y$ as follows:

$$F = \{<\gamma, <\mathbf{ap}(F_1, [\mathbf{ap}(G,\gamma)]_1), \mathbf{ap}(F_2, [\mathbf{ap}(G,\gamma)]_2)>> : \gamma <_{\mathbf{or}} \alpha :\}.$$

Clearly, if G, F_1, and F_2 are 1–1, then F is also 1–1. □

EXAMPLE A.3 Assume that the sets X and Y are α-countable, where α is an ordinal such that $\{0,1\} \times \alpha$ is α-countable. We take G to be an α-enumeration of $\{0,1\} \times \alpha$, F_1 an α-enumeration of X, and F_2 an α-enumeration of Y. We set F as the function with $\mathbf{do}(F) = \alpha$ such that whenever $\gamma <_{\mathbf{or}} \alpha$, then

$$\mathbf{ap}(F,\gamma) = [[\mathbf{ap}(G,\gamma)]_1 = 0 \Rrightarrow \mathbf{ap}(F_1, [\mathbf{ap}(G,\gamma)]_2), \mathbf{ap}(F_2, [\mathbf{ap}(G,\gamma)]_2)].$$

Now, if $Z \in X$ such that $\mathbf{ap}(F_1,\gamma) = Z$, and $\gamma' <_{\mathbf{or}} \alpha$ such that $\mathbf{ap}(G,\gamma') = <0,\gamma)>$, it follows that $\mathbf{ap}(F,\gamma') = Z$. On the other hand, if $Z \in Y$ we can choose $\gamma' <_{\mathbf{or}} \alpha$ such that $\mathbf{ap}(F,\gamma') = Z$. This means that F is an α-enumeration of $X \cup Y$. □

EXERCISE A.12 Assume α is infinite and $\alpha \times \alpha$ is α-countable. Prove that $\{0,1\} \times \alpha$ is α-countable.

The preceding example is rather simple, because we deal with the union of two sets. To deal with the general union operation we must use some form of the axioms of choice. In the next theorem we prove the general case using the completeness axiom from the enumeration operation.

Theorem A.7 *Assume that the set X is α-countable, every element of X is α-countable and, furthermore, $\alpha \times \alpha$ is α-countable. Then $\bigcup X$ is α-countable.*

PROOF. The proof uses the completeness of the enumeration rule. We assume that X is non-empty and also that $\emptyset \notin X$. Assume $F' \in \Upsilon(X,\alpha)$ is an

α-enumeration of X, and $F'' \in \Upsilon(\alpha \times \alpha, \alpha)$ is an α-enumeration of $\alpha \times \alpha$. Let Z be the set,

$$Z = \{F \in \Upsilon(Y, \alpha) : Y \in X : \mathbf{ra}(F) = Y\},$$

and $G \in \Upsilon(Z, \alpha)$ be an α-enumeration induced by the predicate

$$p(V, Y) \equiv Y \in \Upsilon(\mathbf{ap}(F', \mathbf{do}(V)), \alpha).$$

From the assumptions it follows that for every $\gamma <_{\mathbf{or}} \alpha$, $\mathbf{ap}(G, \gamma)$ is an α-enumeration of $\mathbf{ap}(F', \gamma)$. Hence we take $F \in \Upsilon(\bigcup X, \alpha)$ such that

$$\mathbf{ap}(F, \gamma) = \mathbf{ap}(\mathbf{ap}(G, [\mathbf{ap}(F'', \gamma)]_1), [\mathbf{ap}(F'', \gamma)]_2).$$

Clearly, F is an α-enumeration of $\bigcup X$. □

Remark A.8 Theorem A.7 is a critical tool in proofs of countability. For example, assume $X = \{h(Y) : Y \in Z :\}$, where Z is α-countable with α-enumeration F. Then $(\lambda\beta <_{\mathbf{or}} \alpha)h(\mathbf{ap}(F, \beta))$ is an α-enumeration of X. If we know that $h(Y)$ is α-countable whenever $Y \in Z$, it follows that $\bigcup X$ is α-countable. □

EXERCISE A.13 Prove that ω is $(\omega +_{\mathbf{or}} \omega)$-denumerable.

Theorem A.8 *If Z is α-countable, then there exists $\alpha' \leq_{\mathbf{or}} \alpha$ such that Z is α'-denumerable.*

PROOF. Assume that F is a function such that $\mathbf{do}(F) = \alpha$ and $\mathbf{ra}(F) = Z$. Let f be the set operation defined by ordinal recursion, such that

$$f(\gamma) = \mathbf{ap}(F, (\mu\delta <_{\mathbf{or}} \alpha)(\gamma \leq_{\mathbf{or}} \delta \wedge (\forall\delta' <_{\mathbf{or}} \gamma)f(\delta') \neq \mathbf{ap}(F, \delta))).$$

First, we note that given $\delta <_{\mathbf{or}} \alpha$ there exists $\gamma <_{\mathbf{or}} \alpha$ such that $\mathbf{ap}(F, \delta) = f(\gamma)$. This is obvious if there is $\delta' <_{\mathbf{or}} \delta$ such that $\mathbf{ap}(F, \delta) = f(\delta')$, so we take $\gamma = \delta'$. Otherwise, we take $\gamma = \delta$. We define the ordinal α' as follows:

$$\alpha' = (\mu\alpha' \leq_{\mathbf{or}} \alpha)(\forall\delta <_{\mathbf{or}} \alpha)(\exists\gamma <_{\mathbf{or}} \alpha')\mathbf{ap}(F, \delta) = f(\gamma).$$

From the minimality of α' it follows that whenever $\gamma <_{\mathbf{or}} \alpha'$ there exists $\delta <_{\mathbf{or}} \alpha$ such that $\mathbf{ap}(F, \delta) = f(\gamma)$, and hence $f(\gamma) \in Z$. Furthermore, from the definition of f it follows that if $\gamma' <_{\mathbf{or}} \gamma <_{\mathbf{or}} \alpha'$, then $f(\gamma') \neq f(\gamma)$. We conclude that if $F' = (\lambda\gamma <_{\mathbf{or}} \alpha')f(\gamma)$, then F' is 1–1, $\mathbf{do}(F') = \alpha'$, and $\mathbf{ra}(F') = Z$, so Z is α'-denumerable. □

Corollary A.8.1 *If β is α-countable, and $\alpha \leq_{\mathbf{or}} \beta$, then β is α-denumerable.*

PROOF. From Theorem A.8 it follows that there exists $\alpha' \leq_{\mathbf{or}} \alpha$ such that β is α'-denumerable. Using inversion we conclude there is a 1–1 function F such that $\mathbf{do}(F) = \beta$ and $\mathbf{ra}(F) = \alpha' \leq_{\mathbf{or}} \alpha \leq_{\mathbf{or}} \beta$. Using the Cantor-Berstein Theorem (Example 7.2) we get a 1–1 function F' such that $\mathbf{do}(F') = \beta$ and $\mathbf{ra}(F') = \alpha$. Using inversion again, we conclude that β is α-denumerable. □

Corollary A.8.2 *If β and α are ordinals such that β is α-countable and α is β-countable, then β is α-denumerable.*

PROOF. Immediate from Corollary A.8.1, for either $\beta \leq_{\mathbf{or}} \alpha$ or $\alpha \leq_{\mathbf{or}} \beta$. □

EXERCISE A.14 Assume that $X \subseteq Z$ where X is α-denumerable and Z is α-countable. Prove that Z is α-denumerable.

A.2.4 $\mathbf{ini}(\beta) = \mathbf{inf}(\{\alpha \leq_{\mathbf{or}} \beta :: \mathbf{CT}(\beta, \alpha)\})$ (*the least ordinal α such that β is α-countable*).

A.2.5 $\mathbf{IN}(\beta) \equiv (\forall \alpha <_{\mathbf{or}} \beta)\neg\mathbf{CT}(\beta, \alpha)$ (*β is an initial ordinal*).

EXAMPLE A.4 The ordinal ω is initial. For, if $\alpha <_{\mathbf{or}} \omega$ and $\mathbf{CT}(\omega, \alpha)$, then there is a natural number n such that ω is n-denumerable, and by Theorem A.4 this means that ω is finite. □

Theorem A.9 *Let β and α be ordinals. The following conditions are equivalent:*

(i) $\mathbf{ini}(\beta) = \alpha$.

(ii) $\alpha \leq_{\mathbf{or}} \beta \wedge \mathbf{CT}(\beta, \alpha) \wedge (\forall \gamma <_{\mathbf{or}} \alpha) \neg \mathbf{CT}(\beta, \gamma)$.

(iii) $\mathbf{IN}(\alpha) \wedge \mathbf{DE}(\beta, \alpha)$.

PROOF. The implication from (i) to (ii) is clear from definition A.2.4, noting that $\mathbf{CT}(\beta, \beta)$ holds. From (ii) it is clear that $\mathbf{IN}(\alpha)$. To prove $\mathbf{DE}(\beta, \alpha)$, note that β is α-countable, and that it follows from Theorem A.8 that there exists $\alpha' \leq_{\mathbf{or}} \alpha$ such that β if α'-denumerable. Hence from part (ii) it follows that $\alpha' = \alpha$, so $\mathbf{DE}(\beta, \alpha)$. Finally, we assume (iii) to prove (i). From $\mathbf{DE}(\beta, \alpha)$ we derive $\mathbf{CT}(\beta, \alpha)$ and also $\mathbf{DE}(\alpha, \beta)$, hence from $\mathbf{IN}(\alpha)$ we get $\alpha \leq_{\mathbf{or}} \beta$ and $(\forall \gamma <_{\mathbf{or}} \alpha) \neg \mathbf{CT}(\alpha, \beta)$. We conclude that $\mathbf{ini}(\beta) = \alpha$. □

Corollary A.9.1 *If β is an arbitrary ordinal, then* $\mathbf{IN}(\mathbf{ini}(\beta))$.

PROOF. From Theorem A.9, setting $\alpha = \mathbf{ini}(\beta)$. □

EXAMPLE A.5 From Corollary A.9.1 we know that $\mathbf{IN}(\mathbf{ini}(\beta))$ holds for any ordinal β. Furthermore, $\mathbf{DE}(\mathbf{ini}(\beta), \mathbf{ini}(\beta))$ holds trivially. Hence, if we apply Theorem A.9 with $\mathbf{ini}(\beta)$ for β and α we get $\mathbf{ini}(\mathbf{ini}(\beta)) = \mathbf{ini}(\beta)$. □

EXERCISE A.15 Let β and α be ordinals. Prove that the following conditions are equivalent:

(a) β is α-denumerable.

(b) $\mathbf{ini}(\beta) = \mathbf{ini}(\alpha)$.

(c) α is β-denumerable.

EXERCISE A.16 Prove that if $\beta \leq_{\mathbf{or}} \alpha$, then $\mathbf{ini}(\beta) \leq_{\mathbf{or}} \mathbf{ini}(\alpha)$.

EXERCISE A.17 Let β be an ordinal. Prove that $\mathbf{IN}(\beta) \equiv \mathbf{ini}(\beta) = \beta$.

EXERCISE A.18 Let f be a set operation introduced by replacement in the form: $f(Z) = \{h(Y) : Y \in g(Z) : p(Y, Z)\}$. Prove that whenever $g(Z)$ is α-countable, then $f(Z)$ is also α-countable.

A.2.6 $\mathbf{cwe}(\alpha) = \{\mathbf{rk}^{**}(R) : R \in \mathbf{cpw}(\alpha \times \alpha, \alpha) : \mathbf{fd}(R) = \alpha \wedge \mathbf{WO}(R)\}$ (*the well-orders of* α).

A.2.7 $\mathbf{nin}(\alpha) = \mathbf{sup}^{+}(\mathbf{cwe}(\alpha))$ (*the next initial ordinal after* α).

Note that in case $\alpha \times \alpha$ is α-countable, then $\mathbf{cpw}(\alpha \times \alpha, \alpha)$ is the power-set of $\alpha \times \alpha$ (see Remark A.7).

Theorem A.10 *If R is a well-ordered relation and* $\mathbf{rk}^{**}(R) = \alpha$, *then* $\mathbf{fd}(R)$ *is α-denumerable.*

PROOF. From Remark 5.2 there is a 1–1 function F such that $\mathbf{fd}(R) = \mathbf{do}(F)$ and $\mathbf{ra}(F) = \alpha$. We conclude that $\mathbf{fd}(R)$ is α-denumerable. □

Corollary A.10.1 *Let α and β be ordinals, where $\alpha \times \alpha$ is α-countable. The following conditions are equivalent:*

(i) $\beta \in \mathbf{cwe}(\alpha)$.

(ii) *β is α-denumerable.*

PROOF. Assume that (i) holds and that R is a relation that satisfies definition A.2.6. From Theorem A.10 it follows that β is α-denumerable. If (ii) holds, let F be a 1–1 function such that $\mathbf{do}(F) = \alpha$ and $\mathbf{ra}(F) = \beta$. Using the construction in Example 6.4 with $X = \alpha$, we get a relation R that satisfies definition A.2.6, so $\beta \in \mathbf{cwe}(\alpha)$. □

Corollary A.10.2 *Let α and β be ordinals, where $\alpha \times \alpha$ is α-countable. The following conditions are equivalent:*

(i) $\beta <_{\mathbf{or}} \mathbf{nin}(\alpha)$.

(ii) *β is α-countable.*

PROOF. If $\beta <_{\mathbf{or}} \mathbf{nin}(\alpha)$, then $\beta \leq_{\mathbf{or}} \gamma$, where $\gamma \in \mathbf{cwe}(\alpha)$. It follows that β is γ-countable, and by Corollary A.10.1 that β is α-countable. To prove the converse, assume that β is α-countable, and to get a contradiction, assume $\mathbf{nin}(\alpha) \leq_{\mathbf{or}} \beta$. Since $\alpha <_{\mathbf{or}} \mathbf{nin}(\alpha)$, it follows that α is β-countable. Hence, by Corollary A.8.1 it follows that β is α-denumerable, hence $\beta \in \mathbf{cwe}(\alpha)$ and $\beta <_{\mathbf{or}} \mathbf{nin}(\alpha)$ by Definition A.2.7. This is a contradiction, so we conclude that $\beta <_{\mathbf{or}} \mathbf{nin}(\alpha)$. □

EXERCISE A.19 Prove that if $\alpha \leq_{\mathbf{or}} \beta$, then $\mathbf{nin}(\alpha) \leq_{\mathbf{or}} \mathbf{nin}(\beta)$.

Corollary A.10.3 *Let α and β be ordinals, where $\alpha \times \alpha$ is α-countable. The following conditions are equivalent:*

(i) $\beta \in \mathbf{cwe}(\alpha)$.

(ii) $\mathbf{ini}(\alpha) \leq_{\mathbf{or}} \beta <_{\mathbf{or}} \mathbf{nin}(\alpha)$.

PROOF. If (i) holds we know by Corollary A.10.1 that β is α-denumerable, hence α is β-denumerable. We conclude that $\mathbf{ini}(\alpha) \leq_{\mathbf{or}} \beta$. On the other hand, from Corollary A.10.2 it follows that $\beta <_{\mathbf{or}} \mathbf{nin}(\alpha)$. If (ii) holds, then from Corollary A.10.2 we have that β is α-countable, and also α is β-countable, so β is α-denumerable, hence by Corollary A.8.2 β is α-denumerable, and by Corollary A.10.1, $\beta \in \mathbf{cwe}(\alpha)$. □

Theorem A.11 *If X is a set where all the elements are initial ordinals, then $\bigcup X$ is an initial ordinal.*

PROOF. We set $\beta = \bigcup X$ and assume $\beta \notin X$. To get a contradiction, assume that β is not initial, so there exists $\alpha <_{\mathbf{or}} \beta$ such that β is α-countable. This means that there is $\beta' \in X$, where $\alpha <_{\mathbf{or}} \beta' <_{\mathbf{or}} \beta$. It follows that β' is α-countable, contradicting that β' is initial. We conclude that the ordinal β is initial. □

Corollary A.11.1 *If α is an ordinal such that $\alpha \times \alpha$ is α-countable, then* $\mathbf{nin}(\alpha)$ *is an initial ordinal.*

PROOF. We set $\beta = \mathbf{nin}(\alpha)$ and to get a contradiction we assume that β is α'-countable, where $\alpha' <_{\mathbf{or}} \beta$. It follows that α' is α-countable, hence β is α-countable, so by Corollary A.10.2 $\beta <_{\mathbf{or}} \beta$, and this is a contradiction. □

EXERCISE A.20 Prove that if $\alpha <_{\mathbf{or}} \beta$, where β is an initial ordinal, then $\mathbf{nin}(\alpha) \leq_{\mathbf{or}} \beta$.

Corollary A.11.2 *Let F be an ordinal net where $\mathbf{do}(F) = \mathbf{nin}(\alpha)$, $\alpha \times \alpha$ is α-countable, and Z is an α-countable set such that $Z \subseteq \bigcup \mathbf{ra}(F)$. Then there exists $\beta <_{\mathbf{or}} \mathbf{nin}(\alpha)$ such that $Z \subseteq \mathbf{ap}(F, \beta)$.*

PROOF. We introduce a function F' with $\mathbf{do}(F') = Z$, such that $\mathbf{ap}(F', Y) = (\mu\gamma <_{\mathbf{or}} \mathbf{nin}(\alpha))Y \in \mathbf{ap}(F, \gamma)$. Clearly, $\mathbf{ra}(F') = \beta$ is α-countable and $Z \subseteq \bigcup\{\mathbf{ap}(F, \gamma) : \gamma <_{\mathbf{or}} \beta :\}$, so $Z \subseteq \mathbf{ap}(F, \beta)$. □

Remark A.9 We know that ω is the minimal infinite initial ordinal, also that $\omega_1 = \mathbf{nin}(\omega)$ is the next initial ordinal after ω. Still, we cannot show that $\mathbf{nin}(\omega_1)$ is an initial ordinal, because we do not know that ω_1 satisfies the condition: $\omega_1 \times \omega_1$ is ω_1-countable. This matter will be settled in the next section, where we show that ω_1 is a critical ordinal. □

Theorem A.12 *If α and α' are ordinals such that $\alpha \times \alpha$ is α-countable, and α' is α-denumerable, then $\alpha' \times \alpha'$ is α'-countable.*

PROOF. By definition there is a 1–1 function F such that $\mathbf{do}(F) = \alpha$ and $\mathbf{ra}(F) = \alpha'$, and a function F' such that $\mathbf{do}(F') = \alpha$ and $\mathbf{ra}(F') = \alpha \times \alpha$. By composition and inversion we can obtain a function G such that $\mathbf{do}(G) = \alpha'$ and $\mathbf{ra}(G) = \alpha' \times \alpha'$. □

EXERCISE A.21 Prove that if $\alpha' \in \mathbf{cwe}(\alpha)$, then $\mathbf{cwe}(\alpha') = \mathbf{cwe}(\alpha)$.

Theorem A.13 *The set* **hf** *is* ω*-denumerable.*

PROOF. From Definition 7.3.1 it follows that **hf** is the union of an ω-countable set, where the elements are finite sets, so they are also ω-countable. From Theorem A.7 it follows that **hf** is ω-countable. From Theorem A.8 it follows that there is an ordinal $\beta \leq_{\mathbf{or}} \omega$ such that **hf** is β-denumerable. Since **hf** is infinite, we have $\beta = \omega$. □

A.3 Hereditarily Countable Sets

In Chapter 4 we considered the inductive definition of the hereditarily finite sets and showed that the definition can be generalized to an arbitrary predicate p in such a way that Theorem 4.8 and Corollary 4.8.1 are valid in the general setting. Here we apply this technique to define the hereditarily countable sets.

In this section we assume that α is a fixed infinite ordinal such that $\alpha \times \alpha$ is α-countable (for example, ω satisfies these conditions). We also write $\alpha_1 = \mathbf{nin}(\alpha)$. Since α is fixed we consider the predicate **CT** to be a unary predicate, with the second argument equal to α, and write $\mathbf{CT}_\alpha(Z)$ for $\mathbf{CT}(Z, \alpha)$. In the same vein we write $\mathbf{cpw}_\alpha(Z)$ for $\mathbf{cpw}(Z, \alpha)$. Using this notation we introduce by set induction the unary predicate $\mathbf{HC}_\alpha$ as follows:

A.3.1 $\mathbf{HC}_\alpha(Z) \equiv \mathbf{CT}_\alpha(Z) \wedge (\forall Y \in Z)\mathbf{HC}_\alpha(Y)$ *(Z is hereditarily α-countable).*

The inductive definition of $\mathbf{HC}_\alpha$ is of the general form introduced in Definition 4.1.7 with $p(Z) \equiv \mathbf{CT}_\alpha(Z)$, so $\mathbf{HC}_\alpha(Z) \equiv \mathbf{CT}^\dagger_\alpha(Z)$. It follows that Theorem 4.8 and Corollary 4.8.1 apply with this notation. As usual, this definition induces a method of proof by $\mathbf{HC}_\alpha$-induction, and also a rule of $\mathbf{HC}_\alpha$-recursion.

EXERCISE A.22 Assume that Z is an arbitrary set. Prove that the following conditions are equivalent:

(i) $\mathbf{HC}_\alpha(Z)$.

(ii) $\mathbf{WF}(Z) \wedge \mathbf{CT}_\alpha(Z) \wedge (\forall Y \in \mathbf{tc}(Z))\mathbf{CT}_\alpha(Y)$.

EXERCISE A.23 Prove that if Z is hereditarily finite, then Z is hereditarily α-countable.

EXERCISE A.24 Prove that if β is α-countable, then β is hereditarily α-countable.

Theorem A.14 *Let Z be hereditarily α-countable and Y a set. Then:*

(i) *If $Y \subseteq Z$, then Y is hereditarily α-countable.*

(ii) *$Z \setminus Y$ is hereditarily α-countable.*

(iii) *$\bigcap Z$ is hereditarily α-countable.*

(iv) *$\bigcup Z$ is hereditarily α-countable.*

(v) *If Y is hereditarily α-countable, then $Z \cup Y$ is hereditarily α-countable.*

(vi) *$[Z]_1$ and $[Z]_2$ are hereditarily α-countable.*

PROOF. Part (i) follows from Theorem A.3 (i) and Definition A.3.1. Part (ii) follows from (i) noting that $Z \setminus Y \subseteq Z$. To prove (iii), note that either $Z = \emptyset$ or there exists $Z' \in Z$ where Z' is hereditarily α-countable and $\bigcap Z \subseteq Z'$. Part (iv) follows from Theorem A.7. Part (v) follows from (iv), noting that $Z \cup Y = \bigcup\{Z, Y\}$. Part (vi) follows from Definitions 2.3.1 and 2.3.2. □

Corollary A.14.1 *Assume Z and Z' are hereditarily α-countable. Then:*

(i) *$<Z, Z'>$ is hereditarily α-countable.*

(ii) *$Z \times Z'$ is hereditarily α-countable.*

PROOF. Part (i) is clear from Definition 1.6.3. Part (ii) follows from part (i) and Theorem A.6. □

Theorem A.15 *Assume that Z is a set such that $Z \subseteq \mathbf{cpw}_\alpha(Z)$. If $X \in Z$ and X is well-founded, then X is hereditarily α-countable.*

PROOF. With Z fixed, let $p(X) \equiv X \in Z \to \mathbf{HC}_\alpha(X)$. We prove by **WF**-induction that $p(X)$ holds whenever X is a well-founded set. To prove $p(X)$, assume that $X \in Z$, hence X is α-countable and $X \subseteq Z$. If $V \in X$, then $p(V)$ holds by the induction hypothesis and $V \in Z$, hence V is hereditarily α-countable. It follows that X is hereditarily α-countable. □

Corollary A.15.1 *Let X and Z be sets such that $\mathbf{cpw}_\alpha(Z) = Z$. Then the following conditions are equivalent:*

(i) $X \in Z \wedge \mathbf{WF}(X)$.

(ii) $\mathbf{HC}_\alpha(X)$.

PROOF. The implication from (i) to (ii) follows from Theorem A.15. To prove the implication from (ii) to (i) we note that $\mathbf{HC}_\alpha(X)$ implies $\mathbf{WF}(X)$. We use $\mathbf{HC}_\alpha$-induction to prove that $X \in Z$. By definition A.3.1 X is α-countable. If $V \in X$, then V is hereditarily α-countable and $V \in Z$ by the induction hypothesis. This means $X \subseteq Z$, hence $X \in \mathbf{cpw}_\alpha(Z) = Z$. □

Corollary A.15.2 *Let X and Z be sets such that $\mathbf{cpw}_\alpha(Z) = Z$ and Z is well-founded. The following conditions are equivalent:*

(i) $X \in Z$.

(ii) $\mathbf{HC}_\alpha(Z)$.

PROOF. The implication from (i) to (ii) follows from Corollary A.15.1, noting that if $X \in Z$, then X is well-founded. The implication from (ii) to (i) is immediate from Corollary A.15.1. □

Theorem A.16 $\mathbf{cpw}_\alpha^\odot(\alpha_1, \emptyset)$ *is a fixed point of* $\mathbf{cpw}_\alpha$.

PROOF. From Corollary 7.10.1 it follows that we need only to prove that $\mathbf{cpw}_\alpha(\bigcup \mathbf{ra}(F)) \subseteq \bigcup \mathbf{ra}(F)$, where F is the ω_1-net induced by $\mathbf{cpw}_\alpha$ at $\emptyset$. If $Z \in \mathbf{cpw}_\alpha(\bigcup \mathbf{ra}(F))$, then Z is α-countable and $Z \subseteq \bigcup \mathbf{ra}(F)$. By Corollary A.11.2 it follows that there exists $\beta <_{\mathbf{or}} \alpha_1$ such that $Z \subseteq \mathbf{cpw}_\alpha^\odot(\beta, \emptyset)$. It follows that $Z \in \mathbf{cpw}_\alpha^\odot(S(\beta), \emptyset)$, hence $Z \in \mathbf{cpw}_\alpha^\odot(\alpha_1, \emptyset)$. □

From Corollary A.15.2 it follows that $\mathbf{cpw}_\alpha^\odot$ contains exactly the hereditarily α-countable sets. We use this property to define formally:

A.3.2 $\mathbf{hc}_\alpha = \mathbf{cpw}_\alpha^\odot(\alpha_1, \emptyset)$ (*the set of all hereditarily α-countable sets*).

Theorem A.17 *If p is a binary general predicate, then p is local in* $\mathbf{hc}_\alpha$.

PROOF. Assume that Z is α-countable and $(\forall Y \in Z)(\exists V \in \mathbf{hc}_\alpha)p(Y, V)$. Let F be an α-enumeration of Z, and consider the following predicate p':

$$p'(U, V) \equiv p(\mathbf{ap}(F, \mathbf{do}(U)), V).$$

Let F' be an α-enumeration of $\mathbf{hc}_\alpha$ induced by the predicate p'. We set $W = \mathbf{ra}(F')$. It follows that W is hereditarily α-countable, so $W \in \mathbf{hc}_\alpha$. Furthermore, $(\forall Y \in Z)(\exists V \in W)p(Y, V)$. □

Theorem A.18 *Assume that the set operation f is introduced by the replacement rule, and the set* $\mathbf{hc}_\alpha$ *is closed under the given operations in the rule. Then* $\mathbf{hc}_\alpha$ *is closed under f.*

PROOF. We show this in one example, which can be generalized to any application of the rule (see also the proof of Theorem 7.16). Assume f is a unary set operation introduced in the form:

$$f(X) = \{Z \in h(X, Y_1, Y_2) : Y_1 \in g_1(X) \wedge Y_2 \in g_2(X) : p(Z, Y_1, Y_2, X)\},$$

where we assume $\mathbf{hc}_\alpha$ is closed under h, g_1, g_2, and we must show that $f(X)$ is hereditarily α-countable whenever X is hereditarily α-countable. From the definition it follows that

$$f(X) \subseteq \bigcup\{h([Y]_1, [Y]_2, X) : Y \in g_1(X) \times g_2(X) :\}.$$

We note that $g_1(X) \times g_2(X) \in \mathbf{hc}_\alpha$ by Corollary A.14.1, and that $h([Y]_1, [Y]_2, X) \in \mathbf{hc}_\alpha$ whenever $Y, X \in \mathbf{hc}_\alpha$. Hence it follows that $f(X)$ is a subset of an hereditarily α-countable set, and so $f(X)$ is hereditarily α-countable. □

Corollary A.18.1 *The set* $\mathbf{hc}_\alpha$ *is a local universe.*

PROOF. Clearly, the set $\mathbf{hc}_\alpha$ is well-founded and transitive. Any general predicate is local in $\mathbf{hc}_\alpha$, by Theorem A.17. To prove closure under elementary operations we note that $\mathbf{hc}_\alpha$ is closed under the insertion rule, and by Theorem A.18 it is closed under any operation introduced by the replacement rule. We conclude that $\mathbf{hc}_\alpha$ is a local universe. □

Corollary A.18.2 *The ordinal* α_1 *is a critical ordinal.*

PROOF. We know that $\mathtt{U} = \mathbf{hc}_\alpha$ is a local universe and, clearly $\kappa_\mathtt{U} = \alpha_1$, so by Theorem 6.21 $\kappa_\mathtt{U}$ is a critical ordinal. □

From now on we abandon the assumption that the ordinal α is an infinite ordinal such that $\alpha \times \alpha$ is α-countable.

Theorem A.19 *If* α *is an infinite ordinal, then* $\alpha \times \alpha$ *is* α*-denumerable.*

PROOF. By ordinal induction on α. We know the property holds when α is a critical ordinal (Theorem 7.6) and ω is a critical ordinal (Theorem 7.4). So we assume that $\omega <_{\mathbf{or}} \alpha$ and set $\beta = \bigcup\{\gamma <_{\mathbf{or}} \alpha :: \mathbf{CR}(\gamma)\}$. From Corollary 6.21.1 it follows that β is critical. Furthermore, $\mathbf{nin}(\beta)$ is also critical, and $\beta \leq_{\mathbf{or}} \alpha <_{\mathbf{or}} \mathbf{nin}(\beta)$. By Corollary A.10.2 it follows that α is β-countable, and by Corollary A.8.1 α is β-denumerable. We conclude that $\alpha \times \alpha$ is α-denumerable. □

We are now in a position to generate the initial ordinals by ordinal recursion. We use here a special notation where ω_α denotes the *αth* initial ordinal. We consider only the infinite initial ordinals, so we must have $\omega_0 = \omega$. The definition is by primitive ordinal recursion.

A.3.3-1 $\omega_0 = \omega$.

A.3.3-2 $\omega_{S(\alpha)} = \mathbf{nin}(\omega_\alpha)$.

A.3.3-3 $\omega_\alpha = \mathbf{sup}(\{\omega_\gamma : \gamma <_{\mathbf{or}} \alpha :\})$ (α a limit ordinal).

Theorem A.20 *Let α be an ordinal. Then:*

(i) *$\omega \leq_{\mathbf{or}} \omega_\alpha$, and ω_α is an infinite initial ordinal.*

(ii) *If $\gamma <_{\mathbf{or}} \alpha$, then $\omega_\gamma <_{\mathbf{or}} \omega_\alpha$.*

(iii) *$\alpha \leq_{\mathbf{or}} \omega_\alpha$.*

PROOF. Part (i) is clear if $\alpha = 0$. Otherwise, it follows from Definition A.3.2, using Corollary A.10.3 and Theorem A.11, and noting that $\omega <_{\mathbf{or}} \mathbf{nin}(\omega)$. Part (ii) follows by ordinal induction on α, noting that from Definition A.3.3-2 we have $\omega_\alpha <_{\mathbf{or}} \omega_{S(\alpha)}$. Part (iii) follows by ordinal induction on α, noting that by the induction hypothesis, whenever $\gamma <_{\mathbf{or}} \alpha$, then $\gamma \leq_{\mathbf{or}} \omega_\gamma <_{\mathbf{or}} \mathbf{nin}(\omega_\gamma) \leq_{\mathbf{or}} \omega_\alpha$, hence $\alpha \leq_{\mathbf{or}} \omega_\alpha$. □

Theorem A.21 *Let β be an infinite initial ordinal. Then there is an ordinal $\alpha \leq_{\mathbf{or}} \beta$ such that $\omega_\alpha = \beta$.*

PROOF. The proof is by ordinal induction on β. The case $\beta = \omega$ is trivial, so we assume $\omega <_{\mathbf{or}} \beta$. We set $\alpha = \{\gamma : \gamma <_{\mathbf{or}} \beta : \omega_\gamma <_{\mathbf{or}} \beta\}$ and, clearly, α is an ordinal such that $0 <_{\mathbf{or}} \alpha \leq_{\mathbf{or}} \beta$. We shall prove that $\omega_\alpha = \beta$.

This is clear if α is a successor ordinal, so we assume that α is a limit ordinal. If $\gamma <_{\mathbf{or}} \alpha$, then $\omega_\gamma <_{\mathbf{or}} \beta$, so from the definition of ω_α it follows that $\omega_\alpha \leq_{\mathbf{or}} \beta$. To prove $\beta \leq_{\mathbf{or}} \omega_\alpha$, assume that $\delta <_{\mathbf{or}} \beta$. Since β is initial, from the induction hypothesis it follows that $\mathbf{ini}(\delta) = \omega_\gamma$, where $\gamma \leq_{\mathbf{or}} \omega_\gamma = \mathbf{ini}(\delta) \leq_{\mathbf{or}} \delta <_{\mathbf{or}} \omega_{S(\gamma)} \leq_{\mathbf{or}} \beta$. From the definition of α it follows that $\gamma <_{\mathbf{or}} \alpha$, hence $\delta <_{\mathbf{or}} \omega_{S(\gamma)} \leq_{\mathbf{or}} \omega_\alpha$ We conclude that $\beta \leq_{\mathbf{or}} \omega_\alpha$. □

A.4 Induction Revisited

As an application of the enumeration rule we show how to define explicitly inductive predicates introduced under the set induction rule. For this purpose we need only the operation Υ in the form $\Upsilon(Z, \omega)$. The inductive and recursive properties of ω are essential for this construction.

We return to the predicate ind considered in Section 3.2]. Here again we want to give an explicit definition of the predicate, this time in the system $\mathbb{G}$, using the enumeration rule.

We use primitive recursion to introduce the binary set operation f such that

$$f(0, Z) = \{Z\}.$$

$$f(S(\mathsf{n}), Z) = \bigcup\{\mathbf{is}_{\mathsf{ind}}(V) : V \in f(\mathsf{n}, Z) :\}.$$

The operation f describes the process that generates the inductive tree for Z in Section 3.2]. Next, we define an operation f^* that collects all the nodes occurring in the tree.

$$f^*(Z) = \bigcup\{f(\mathsf{n}, Z) : \mathsf{n} \in \omega :\}.$$

Theorem A.22 *Assume that $V \in f(\mathsf{n}, Z)$. If $V' \in f(\mathsf{m}, V)$, then $V' \in f(\mathsf{n} +_{\mathbf{or}} \mathsf{m}, Z)$.*

PROOF. Immediate by numerical induction on m. □

Corollary A.22.1 *If $V \in f^*(Z)$, then $f^*(V) \subseteq f^*(Z)$.*

PROOF. Immediate from Theorem A.22. □

Corollary A.22.2 *Assume that the induction for* ind *is extensional, so q_1 is false, q_2 is true, and $\mathbf{is}_{\mathsf{ind}}(Z) = g(Z)$. If $\neg\mathsf{ind}(Z)$, then there is a set $V \subseteq f^*(Z)$ such that $Z \in V$, and whenever $Z' \in V$, then $g(Z') \cap V \neq \emptyset$.*

PROOF. With Z fixed, let p be the predicate,

$$p(W,Y) \equiv (W = \emptyset \wedge Y = Z) \vee (W \neq \emptyset \wedge Y \in g(\mathbf{ap}(W, \bigcup \mathbf{do}(W))) \wedge \neg\mathsf{ind}(Y).$$

Let F be an ω-enumeration of $f^*(Z)$ induced by the predicate p. Clearly, $\mathbf{ap}(F, 0) = Z$, and whenever n is an arbitrary natural number, then $\mathbf{ap}(F, S(\mathsf{n})) \in g(\mathbf{ap}(F, \mathsf{n})) \wedge \neg\mathbf{ap}(F, S(\mathsf{n}))$. So we take $V = \mathbf{ra}(F)$. □

Finally, we use the enumeration rule to introduce explicitly the predicate ind' and the set operation $\mathbf{tc}_{\mathsf{ind}'}$.

$rs(F,Z) \equiv F \in \Upsilon(f^*(Z),\omega) \wedge \mathbf{ap}(F,0) = Z \wedge (\forall \mathsf{n})((\mathbf{is}_{\mathsf{ind}}(\mathbf{ap}(F,\mathsf{n})) = \emptyset \wedge \mathbf{ap}(F,S(\mathsf{n})) = \mathbf{ap}(F,\mathsf{n})) \vee (\mathbf{ap}(F,S(\mathsf{n})) \in \mathbf{is}_{\mathsf{ind}}(\mathbf{ap}(F,\mathsf{n}))))$ (*F is a reduction sequence for Z*).

$\mathsf{ind}'(Z) \equiv (\forall F \in \Upsilon(f^*(Z),\omega))(rs(F,Z) \rightarrow (\exists \mathsf{n})(\mathbf{is}_{\mathsf{ind}}(\mathbf{ap}(F,\mathsf{n})) = \emptyset \wedge (q_1(\mathbf{ap}(F,\mathsf{n})) \vee q_2(\mathbf{ap}(F,\mathsf{n}))))).$

$\mathbf{tc}_{\mathsf{ind}'}(Z) = [\mathsf{ind}'(Z) \Rrightarrow \bigcup\{\mathbf{ra}(F) : F \in \Upsilon(f^*(Z),\omega) : rs(F,Z)\}, \emptyset].$

EXAMPLE A.6 Assume F is a reduction sequence for Z and $\mathbf{ap}(F,1) = Y$. We set $F' \in \Upsilon(Y,\omega)$ to be $\mathbf{ap}(F',\mathsf{n}) = \mathbf{ap}(F,S(\mathsf{n}))$. From the definition and the preceding remark it follows that F' is a reduction sequence for Y. □

EXAMPLE A.7 Assume F is a reduction sequence for Y, where $Y \in \mathbf{is}_{\mathsf{ind}}(Z)$, $q_1(Z)$ fails, and $q_2(Z)$ holds. We set $F' \in \Upsilon(f^*(Z),\omega)$ to be $\mathbf{ap}(F',0) = Z$ and $\mathbf{ap}(F',S(\mathsf{n})) = \mathbf{ap}(F,\mathsf{n})$. From the definition it follows that F' is a reduction sequence for Z. □

Remark A.10 A reduction sequence for Z is equivalent to a branch in the inductive tree for Z, as defined in Section 3.2]. The only difference is that a branch may halt at $\mathbf{ap}(F, \mathsf{n}) = Y$, while the reduction sequence will continue to repeat the same value Y forever. With this understanding, the proofs of the axioms in the induction rule can be obtained by a direct translation of the arguments in Section 3.2]. In particular, the branch generated in the proof of the foundation axiom is obtained here as a function $F \in \Upsilon(f^*(Z), \omega)$ induced by the completeness axiom. □

Remark A.11 The preceding construction shows that inductive predicates can be introduced by standard procedures in set theory, and that special axioms are not required. On the other hand, a similar construction does not seem possible for the set recursion rule, at least in the predicative frame in which we are operating here. Note also that the construction requires the inductive and recursive properties of the set ω. Similarly, the ordinal permutation rule involves the inductive and recursive property of the ordinals. In any case, we consider that the set induction and set recursion rules constitute a unit that is being organized into two rules only for convenience. Furthermore, the constructions for these rules are more primitive, as they do not involve special sets. For this reason, we have decided to avoid the illusory economy provided by the ordinal enumeration rule, and to take as primitive the induction and recursion rules. In the same sense, see Remark 3.5 □

A.5 Notes

The ordinal enumeration rule provides a reasonable alternative to the ordinal permutation rule. The latter is open to several criticisms, in particular for potentially invoking the class of all ordinals, and depending on the principle of closure, which amounts to a restriction on the sets in the universe. The ordinal enumeration rule introduces an operation $\Upsilon(Z, \alpha)$ that is completely determined by the arguments Z and α, so it is independent of the universe of sets or the class of ordinals. Some of the applications given above are still simpler, for they require only the ordinal ω.

By replacing the ordinal permutation rule with the ordinal enumeration rule we obtain a weaker sub-system that makes room for a number of important constructions in classical set theory. The distinction between countable and non-countable sets is undoubtedly a major contribution in standard set theory. In the first place, for making explicit a fundamental property of the set of natural numbers and secondly for showing that it is not an intrinsic property of sets, for it fails for some forms of collections.

An intriguing consequence of the ordinal enumeration rule is the explicit definition of inductive predicates, and this makes possible the elimination of the set induction rule in Chapter 3. On the other hand, it does not seem possible to eliminate the rule of set recursion in Chapter 5. The standard proof of transfinite recursion cannot be translated into the system $\mathbb{G}$, for it requires an argument by induction with a non-local predicate.

Under the ordinal enumeration rule the power-set construction is allowed only for well-ordered sets, and cannot be iterated. Still, cardinals can be defined via initial ordinals. Furthermore, a weak version of the axiom of choice is available.

Bibliography

Aczel, P. (1988). *Non-well-founded sets.* Stanford: Center for the Study of Language and Information.

Barwise, J. (1975). *Admissible Sets and Structures.* Berlin: Springer.

Boolos, G. (1971) "The iterative concept of set." *The Journal of Philosophy*, 68:215–31.

Cantor, G. (1895). "Beiträge zur Begründung der transfiniten Mengelehre I." *Math. Ann.* 46:481–512

Cavaillès, J. (1962). *Philosophie Mathématique.* Paris:Hermann.

Curry, H. B. (1963). *Foundations of Mathematical Logic.* New York:McGraw-Hill Book Company,Inc.

Drake, F. R. (1974). *Set Theory, An Introduction to Large Cardinals.* Amsterdam: North-H lland.

Enderton, H. B. (197'). *Elements of Set Theory.* New York:Academic Press.

Gentzen, G. (1934). "Untersuchungen über das logische Schliessen" *Mathematische Zeitschrift* 39:405-431.

Hallett, M. (1984). *Cantorian Set Theory and Limitations of Size.* Oxford: Clarendon Press.

Hinman, P. G. (1978). *Recursion-Theoretic Hierarchies.* Berlin:Springer.

Kleene S. C. (1952). *Introduction to Metamathematics.* Amsterdam: North-Holland Publishing Co.

Kunen, K. (1980). *Set Theory: An Introduction to Independence Proofs.* Amsterdam: North Holland.

Levy, A. (1979). *Basic Set Theory.* Berlin: Springer.

Mirimanoff, D. (1917) "Les antinomies de Russell et de Burali-et le problème fondamental de la théorie des ensembles." *L'Enseignement Math.* 19:37-52.

Montague, R. (1955). "Well-founded relations: generalizations of principles of induction and recursion" (abstract). *Bull. Amer. Math. Soc.* 61:442.

Moschovakis, Y. N. (1991). *Notes on Set Theory.* New York: Springer-Verlag.

Mostowski, A. (1949). "An undecidable mathematical statement." *Fund. Math.* 36:143–164.

Myhill, J. (1975). "Constructive set theory." *The Journal of Symbolic Logic.* 40:347–382.

Sanchis, L. (1992). *Recursive Functionals.* Amsterdam: North Holland.

Shoenfield, J. (1967). *Mathematical Logic.* Reading, MA: Addison-Wesley.

Sochor, A. (1984). "The alternative set theory and its approach to Cantor's set theory." *Aspects of Vagueness*, H. J. Skala, S. Termini, E. Trillas, eds. Dordrecht: D. Reidel Publishing Company.

Tarski, A. (1955). "General principles of induction and recursion; the notion of rank in axiomatic set theory and some of its applications" (2 abstracts). *Bull. Amer. Math. Soc.* 61:442–443.

Veblen, O. (1908). "Continuous increasing functions of finite and transfinite ordinals." *Trans. Amer. Math. Soc.* 9:280–292.

von Neumann, J. (1923). "Zur Einführung der transfiniten Zahlen." *Acta Sc. Math.* Szeged I, 199–208.

von Neumann, J. (1928). "Über die Definition durch transfinite Induktion und verwandte Fragen der allgemeinen Mengelehre. *Math. Ann.* 99:373–391.

Weyl, H. (1949). *Philosophy of Mathematics and Natural Science.* Princeton: Princeton University Press.

Whitehead, A. N., Russell, B. (1912). *Principia Mathematica*, Vol. II.

Zermelo, E. (1904). "Beweis dass jede Menge wohlgeordnet werden kan." *Math. Ann.* 59:514–516.

Primitive Rules and Pages Where Introduced

Formal Definitions and Pages Where Introduced

Index